Springer Proceedings in Mathematics & Statistics

Volume 374

This book series features volumes composed of selected contributions from workshops and conferences in all areas of current research in mathematics and statistics, including operation research and optimization. In addition to an overall evaluation of the interest, scientific quality, and timeliness of each proposal at the hands of the publisher, individual contributions are all refereed to the high quality standards of leading journals in the field. Thus, this series provides the research community with well-edited, authoritative reports on developments in the most exciting areas of mathematical and statistical research today.

More information about this series at https://link.springer.com/bookseries/10533

Susana Relvas · João Paulo Almeida ·
José Fernando Oliveira · Alberto Adrego Pinto
Editors

Operational Research

IO 2019, Tomar, Portugal, July 22–24

Springer

Editors
Susana Relvas
Centro de Estudos de Gestão do Instituto
Superior Técnico (CEG-IST)
University of Lisbon
Lisbon, Portugal

João Paulo Almeida
CeDRi
Department of Mathematics
Polytechnic Institute of Bragança
Braganca, Portugal

José Fernando Oliveira
INESC TEC
Faculty of Engineering
University of Porto
Porto, Portugal

Alberto Adrego Pinto
INESC TEC
Department of Mathematics
University of Porto
Porto, Portugal

ISSN 2194-1009 ISSN 2194-1017 (electronic)
Springer Proceedings in Mathematics & Statistics
ISBN 978-3-030-85475-1 ISBN 978-3-030-85476-8 (eBook)
https://doi.org/10.1007/978-3-030-85476-8

Mathematics Subject Classification (2010): 90BXX

This Springer imprint is published by the registered company Springer Nature Switzerland AG
The registered company address is: Gewerbestrasse 11, 6330 Cham, Switzerland

Preface

In 2019, APDIO has organized the twentieth Portuguese national conference of Operational Research (OR)—the IO series. With this long-lasting list of events, success has been the main takeaway from each unique occurrence.

With a growing number of members, also with growing enthusiasm, the conference series has reached most regions in the mainland country. Recently, the frequency of the conference was increased, while maintaining the same attendance and quality of work. Outreach of the conference to OR researchers from different nationalities is also increasing and bringing an international feel to the events.

Year after year, keynote speakers from industry and from other European societies have joined us to share knowledge and current research and practice of OR. Our national OR discussion forum has influenced, as well, a growing number of national members attending and presenting in international OR conferences and participating actively in EURO Working Groups.

However, besides validating our OR research, our national conferences always promote a very strong, very Portuguese social occasion. A true networking event is where work and social well-being are sustained—and looked forward—event after event.

With this book as the last physical milestone of IO2019, we perpetuate the results from this event and invite all our members, and OR researchers out there, to continue this History in the next events.

Lisbon, Portugal — Susana Relvas
Braganca, Portugal — João Paulo Almeida
Porto, Portugal — José Fernando Oliveira
Porto, Portugal — Alberto Adrego Pinto

Contents

Contributors

Ana Raquel Aguiar Centre for Management Studies, Instituto Superior Técnico (CEG-IST), Universidade de Lisboa, Lisbon, Portugal

João P. Almeida CeDRI – Research Centre in Digitalization and Intelligent Robotics, Instituto Politécnico de Bragança, Bragança, Portugal

Ana Paula Barbosa-Póvoa Centre for Management Studies (CEG-IST), Instituto Superior Técnico, University of Lisbon, Lisboa, Portugal

Ana Camanho Faculdade de Engenharia, Universidade do Porto, Porto, Portugal

Ana Carvalho CEG-IST, Instituto Superior Técnico, Universidade de Lisboa, Lisboa, Portugal

Fábio Coelho CEG-IST, Instituto Superior Técnico, Universidade de Lisboa, Lisboa, Portugal

Diego Câmara Centro de Estudos de Gestão-IST, Instituto Superior Técnico, Universidade de Lisboa, Lisboa, Portugal

Cátia da Silva CEG-IST, Instituto Superior Técnico, Universidade de Lisboa, Lisboa, Portugal

Carolina Soares de Morais Centre for Management Studies, Instituto Superior Técnico (CEG-IST), Universidade de Lisboa, Lisbon, Portugal

Luís Pinto Ferreira Centre for Research & Development in Mechanical Engineering (CIDEM), Instituto de Ciência e Inovação em Engenharia Mecânica e Engenharia Industrial (INEGI), School of Engineering of Porto (ISEP), Polytechnic of Porto, Porto, Portugal

Angelo Aliano Filho Federal Technological University of Paraná, Apucarana, Paraná, Brazil

Dorabela Gamboa CIICESI, Escola Superior de Tecnologia e Gestão, Politécnico do Porto, Felgueiras, Portugal

Carla A. S. Geraldes CeDRI – Research Centre in Digitalization and Intelligent Robotics, Instituto Politécnico de Bragança, Bragança, Portugal;
Centro ALGORITMI, Universidade do Minho, Guimarães, Portugal

Bruno Gonçalves Departamento de Produção e Sistemas, Universidade do Minho, Guimarães, Portugal

Joana Guapo CEG-IST, Instituto Superior Técnico Universidade de Lisboa, Lisboa, Portugal

Xenia Klimentova INESC TEC, Campus da FEUP, Porto, Portugal

Cristina Lopes CEOS.PP/ISCAP/P.Porto and LEMA, Polytechnic of Porto, Porto, Portugal

Rodrigo Macedo CEG-IST, Instituto Superior Técnico, Universidade de Lisboa, Lisboa, Portugal

Inês Marques Centre for Management Studies, Instituto Superior Técnico, University of Lisbon, Lisbon, Portugal

Ali Moghanni CMUC, Department of Mathematics, University of Coimbra, Coimbra, Portugal

Samuel Moniz Univ Coimbra, CEMMPRE, Departamento de Engenharia Mecânica, Universidade de Coimbra, Coimbra, Portugal

Tiago Monteiro INESC TEC, Campus da FEUP, Porto, Portugal

Bruna Mota CEG-IST, Instituto Superior Técnico, Universidade de Lisboa, Lisboa, Portugal

Mariana Oliveira Centre for Management Studies, Instituto Superior Técnico, University of Lisbon, Lisbon, Portugal

Silvana Oliveira School of Engineering of Porto (ISEP), Polytechnic of Porto, Porto, Portugal

Óscar Oliveira CIICESI, Escola Superior de Tecnologia e Gestão, Politécnico do Porto, Felgueiras, Portugal

Marta Pascoal CMUC, Department of Mathematics, University of Coimbra, Coimbra, Portugal;
Institute for Systems Engineering and Computers – Coimbra, University of Coimbra, Coimbra, Portugal

João Pedro Pedroso INESC TEC, Campus da FEUP, Porto, Portugal;
Faculdade de Ciências, Universidade do Porto, Porto, Portugal

Tânia Pinto-Varela Centro de Estudos de Gestão-IST, Instituto Superior Técnico, Universidade de Lisboa, Lisboa, Portugal

Tânia Rodrigues Pereira Ramos Centre for Management Studies, Instituto Superior Técnico (CEG-IST), Universidade de Lisboa, Lisbon, Portugal

Susana Relvas CEG-IST, Instituto Superior Técnico, Universidade de Lisboa, Lisboa, Portugal

João Pires Ribeiro CEG-IST, Instituto Superior Técnico, Universidade de Lisboa, Lisboa, Portugal

Rogério Rocha Center for Power and Energy Systems, INESC-TEC, Faculdade de Engenharia, Universidade do Porto, Porto, Portugal

Fabiane K. Setti Instituto Politécnico de Bragança, Bragança, Portugal;
Universidade Tecnológica Federal do Paraná, Paraná, Brazil

Elsa Silva INESC TEC, Faculdade de Engenharia, Universidade do Porto, Porto, Portugal

Helenice Florentino Silva Institute of Bioestatistics of Botucatu, State University of São Paulo, São Paulo, Brazil

Micael Simões Center for Power and Energy Systems, INESC-TEC, Faculdade de Engenharia, Universidade do Porto, Porto, Portugal

M. Teresa Pereira Centre for Research & Development in Mechanical Engineering (CIDEM), Instituto de Ciência e Inovação em Engenharia Mecânica e Engenharia Industrial (INEGI), School of Engineering of Porto (ISEP), Polytechnic of Porto, Porto, Portugal

Marcelo G. Trentin Universidade Tecnológica Federal do Paraná, Paraná, Brazil

Ana Viana INESC TEC, Campus da FEUP, Porto, Portugal;
ISEP - School of Enginnering, Polytechnic of Porto, Porto, Portugal

Miguel Vieira CEG-IST, Instituto Superior Técnico Universidade de Lisboa, Lisboa, Portugal;
Univ Coimbra, CEMMPRE, Departamento de Engenharia Mecânica, Universidade de Coimbra, Coimbra, Portugal

Searching for a Solution Method for the Smart Waste Collection Routing Problem

Ana Raquel Aguiar, Carolina Soares de Morais, Tânia Rodrigues Pereira Ramos, and Ana Paula Barbosa-Póvoa

Abstract To optimize waste collection operations, approaches to define dynamic routes based on the use of real-time information transmitted by sensors are being explored—the so-called Smart Waste Collection Routing Problem (SWCRP). In a previous work, this problem was modeled as a Vehicle Routing Problem with Profits (VRPP), where the profit associated with the collection of recyclable waste is maximized. However, despite the obtained gains, the authors dealt with higher computational times and gaps. To overcome this limitation, an optimization-based heuristic is proposed to decompose the SWCRP, improving solutions' quality while reducing computational times. Through a Cluster First-Route Second methodology, the proposed approach selects a dynamic set of waste bins to be considered, and then uses this dynamic set to feed a VRPP model that decides which waste bins are worth to be collected, considering their fill-levels and locations. Computational experiments are performed and results show the potentiality of the proposed method.

Keywords Smart waste collection routing problem · Optimization-based heuristic · Real-time information

1 Introduction

Nowadays, circular economy (CE) is a societal concern that explores sustainability goals through a cyclical growth model, reducing externalities and stimulating the diversification of business models [1]. Such model allows the enhancement of resources productivity, generating more wealth while reducing exposure to resource's price volatility, promoting a more stabilized economy. Evidence suggests that by 2030 Europe could be generating 600 billion euros of annual economic gains, in addition to savings resulting from the reduction of externalities [2]. To achieve this, business

A. R. Aguiar · C. S. de Morais · T. R. P. Ramos (✉) · A. P. Barbosa-Póvoa
Centre for Management Studies, Instituto Superior Técnico (CEG-IST), Universidade de Lisboa, Avenida Rovisco Pais, 1049-001 Lisbon, Portugal
e-mail: tania.p.ramos@tecnico.ulisboa.pt

S. Relvas et al. (eds.), *Operational Research*, Springer Proceedings in Mathematics & Statistics 374, https://doi.org/10.1007/978-3-030-85476-8_1

models need to explore sustainability goals while embracing current technology development as a mean to support these business models challenges.

In Portugal, one of the first business models integrated into CE was the collection of recyclable waste material. Urban waste is collected by municipalities, while the recyclable waste is collected by semi-private companies, where sustainability goals should be pursuit and an efficient waste management operation needs to be in place. However, as highlighted by [3] and [4], in a recyclable collection system, only 40% of the collected waste bins present fill-levels higher than 75%, and around 38% of the total waste bins are empty when collected. This happens because collection operations are based on the performance of pre-defined and static routes, where all waste bins are visited regardless of their actual fill-levels. A solution to deal with these inefficient operations is to reduce the uncertainty on the bin's fill-level. This can be achieved through the installation of sensors inside the waste bins, which will transmit information that can then be used to define dynamic collection routes [5].

This issue has been studied by [4], that introduced the Smart Waste Collection Routing Problem (SWCRP) and proposed a Vehicle Routing Problem with Profits (VRPP) mathematical model that maximizes the profit (defined as the difference between revenues from selling the recyclable waste and the transportation cost) associated with the selective collection of recyclable waste. This selective collection was made based on real-time information on the bins' fill-levels, transmitted by sensors located inside the waste bins. However, despite the promising values of profit, the approach with the best overall profit presented significant computational times and failed to prove optimality by presenting significant gap values. Thus, a need to invest in efficient solution methods for the SWCRP was identified by the authors.

In this paper, we explore this need and propose an optimization-based heuristic for decomposing the SWCRP, aiming to improve the solutions' quality (finding and proving the optimal solution regarding the profit's maximization) while reducing the computational time required.

The remainder of this paper is structured as follows. In Sect. 2, a literature review on the topic is made. Then in Sect. 3 a new solution methodology for the SWCRP is defined, followed by Sect. 4 where a case-study is characterized. In Sect. 5, the results are presented and discussed, and the paper is finished with Sect. 6 where the main work conclusions are made.

2 Literature Review

In a society characterized by increasing concerns towards sustainability goals, waste management plays a very important role. Thus, waste collection is nowadays an increasingly important business, which is still however characterized by an inefficient operation due to the high uncertainty associated with the real waste bins' fill-levels. But the current developments in Information and Communication Technologies (ICT) have been recently explored to improve the processes of operational decision making in waste management, as reported by [6] and [7]. In [6], four of the analyzed articles are considered to target applications of route collection/optimization. One of them, [5] used analytical modeling and discrete-event simulation to evaluate different policies of scheduling and routing, using real-time data, and concluded that, when compared to static policies, dynamic scheduling and routing policies are able to reduce operations cost, distance and labor hours. When considering real-time data to optimize waste collection operations through exact models, [4] proposed the SWCRP: a VRPP that selects to be collected only those waste bins that are worth to, maximizing the amount of collected waste while minimizing the total travelled distance. Despite the obtained gains in profit, the authors dealt with higher computational times and gaps. To overcome this limitation, optimization-based heuristics could be explored to decompose the VRPP mathematical model. As an example of this methodology, [8] proposed a novel three-phase heuristic approach for the multi-depot VRP with time windows and heterogeneous vehicles, derived from embedding a heuristic-based clustering algorithm within a VRP with Time Windows (VRPTW) optimization framework. By efficiently solving small case studies, the authors realized that it would be beneficial to perform a pre-processing stage, clustering nodes so that a more compact cluster-based MILP problem formulation is achieved. This compact model could then be solved resulting in a performance enhancement. This is not the only case as for solving VRPs, optimized-based heuristics have been an option for quite some time. To the authors knowledge, a first approach was introduced by [9], who solved a VRP by initially assigning a set of feasible customers to each vehicle, and then solving a generalized Traveling Salesman Problem (TSP) for each of the sets obtained in the first stage. This approach has become known as "Cluster First—Route Second" and has been widely used for solving large-scale VRPs, being, together with the "Route First—Cluster Second", the methodology among most of the existing heuristic approaches to solve large size VRPs in reasonable computational times [10]. The Route First—Cluster Second was first introduced by [11] and considers first forming a 'giant tour' from the depot around all the customers and back to the depot (i.e. a TSP tour around all the customers including the depot) and then optimally partitioning such a tour into a set of feasible vehicle routes.

When considering Cluster First—Route Second applied to waste collection using real-time data, a recent work by [4] studied a methodology in which the bins to be visited are firstly clustered and then a Capacitated VRP is solved for each day of the considered period, obtaining a negative profit in the global solution. The heuristic considered for selecting the bins is the establishment of a fill-level threshold of 80%. Such a high threshold value implied a selection of a small set of waste bins, which led to the worst results presented by the proposed heuristic. Thus, the authors have identified the need to further study this problem and overcome the identified limitation.

This need is explored in this paper where a new optimization-based heuristic to solve the SWCRP using real-time information is proposed. This new approach is an evolution of the Cluster First—Route Second one proposed by [4] and combines this technique with a latter collection heuristic to solve the SWCRP, where a VRPP is solved to collect only those waste bins that are worth to, instead of solving a Capacitated VRP, where all waste bins must be visited. To avoid the selection of a small set of waste bins to be visited, a fill-level threshold (M), established for the dynamic set of waste bins selection, is explored though a sensitive analysis so as to conclude on its adequate value.

3 A New Solution Methodology for the Smart Waste Collection Routing Problem

3.1 *Problem Statement*

The SWCRP, introduced by [4], can be defined as follows. Given a set of n waste bins equipped with volumetric sensors, the problem consists in determining the waste bins collection sequence that will maximize the profit, departing from and arriving at the same depot. For that purpose, an undirected graph is considered, composed of $n + 1$ nodes, and with distances (d_{ij}), associated to each arc (i, j).

Capacity limitations are considered associated to the vehicle waste capacity, Q (in kg), and each waste bin i has a volume capacity, E_i. The differentiation of the capacity measurement unit (weight and volume) is an adaptation to the data transmitted by the sensors which is, for each waste bin i at each day t, its fill-level as a percentage of its maximum volume, S_i^t. Within the problem's context, it is convenient to convert the waste bins' fill-level into weight, using in this way the same measure to both vehicle capacity and profit estimation. Thus, the density of the recyclable waste (B) must be used to allow the estimation of the weight of waste present in bin i after the sensor's read is obtained, as $w_i^t = E_i B S_i^t$.

The profit is calculated as the difference between the revenues of waste collection and the costs of its transportation. Revenues are defined as the product of the weight collected and the selling price per weight unit of recyclable material (R), while costs are defined as the product of the travelled distance and the transportation cost per distance unit (C). In this problem, no waste bin is allowed to overflow, guaranteeing a 100% service level. In order to gauge the possibility of a waste bin i at day t overflowing, a parameter representing the expected daily filling rate ($\hat{a}_i^t$) is estimated (e.g. based on historical data).

3.2 *Solution Methodology*

As the SWCRP is extremely demanding when applied to real-world size instances and being at stake an operational level decision where routes are intended to be designed and performed as quickly as possible it is imperative to explore efficient solution approaches. One of such approaches is the decomposition of the problem using optimization-based heuristics for which a solution which is better than the current situation can be achieved in a timely manner. The methodology proposed in this work explores this approach and consists of two phases. Firstly, a heuristic is defined to reduce the number of bins to be considered for collection. Then, in the second phase, the resulting subproblem is modeled as a VRPP and solved.

The heuristic is designed considering the problem's objective, where the aim is to increase waste collection profit. Thus, by considering only waste bins with fill-levels above a predetermined threshold (M) for the routing phase, a good solution would be secured. In addition, delaying the collection as much as possible will guarantee a good solution, which is achieved by not performing collection routes on day t if no waste bin is expected to overflow at the next day, $t+1$. To verify the overflow status of a bin i at the following day ($t+1$), an expected maximum fill-level ($\hat{F}_i^t$), in percentage, is estimated at the end of day t. Considering that the sensors' data are collected in the morning, the parameter $\hat{F}_i^t$ can be determined as the sum of the sensors' information (S_i^t) and the expected daily filling rate ($\hat{F}_i^t = S_i^t + \hat{a}_i^t$). However, considering that the collection does not happen immediately after the sensors' transmission and that it is possible that more than the estimated waste exists inside the waste bin, a conservative stance is adopted ($\hat{F}_i^{t+1} = S_i^t + \hat{a}_i^t + \hat{a}_i^{t+1}$). In this sense, the condition to the existence of a route on day t is the existence of at least one bin that is expected to overflow on day $t+1$, i.e. is the existence of at least one waste bin for which $\hat{F}_i^{t+1}$, in percentage of the waste bin's capacity, is equal or greater than 100% ($\hat{F}_i^{t+1} \geq 1$).

After identifying the day to perform collection routes, a dynamic set (L) of waste bins to consider in the routing phase is formed by the waste bins that present fill-levels above a threshold M. A solution is then obtained by solving a VRPP for the dynamic set L. The developed methodology is schematically represented in Fig. 1.

Summarizing, based on the sensors' information received (S_i^t) on the morning of day t it is verified if there is at least one waste bin which will overflow at the next day $t+1$. If it exists, the dynamic set (L) of waste bins to be considered for the routing is defined considering those ones for which the fill-levels overpass a threshold M. For those waste bins, the following VRPP model is solved. Due to its computational efficiency, and as in [4], the two-commodity flow formulation proposed by [12] is used in the model. This formulation considers two flow-variables to define two flow paths for any route of a possible solution: one path from the real depot to the copy depot is given by the flow variables representing the vehicle load, while the second path from the copy depot to the real depot is given by the flow variables representing the empty space in the vehicle.

Sets

$I = 0, 1, 2, \ldots, n+1$: set of n waste bins and the real depot 0 and the copy depot $n+1$

$L \subseteq I$: dynamic subset of waste bins for which $S_i^t > M$

Parameters

C: travelling cost per distance unit (in euro/km)

R: selling price per kg of a recyclable material (in euro/kg)

Q: vehicle weight capacity (in kg)

B: waste density (in kg/m^3)

E_i: volume capacity of bin i (in m^3)

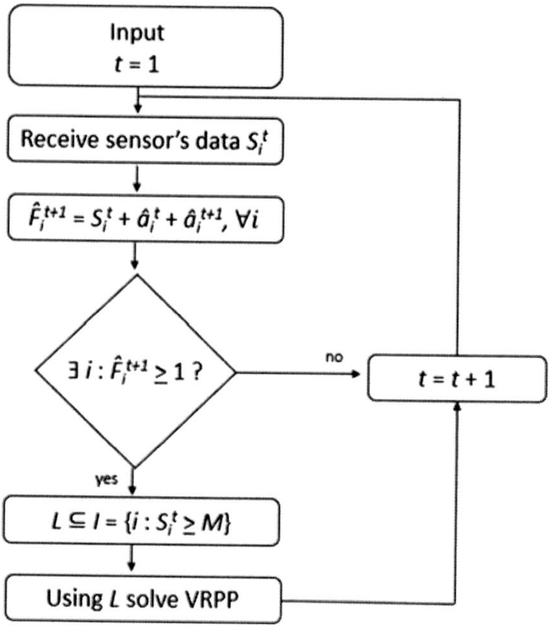

Fig. 1 Schematic representation of the proposed methodology

$E_i B$: weight capacity of bin i (in kg)

d_{ij}: distance between node i and node j (in km)

S_i^t: : fill level in percentage of volume of bin i at the beginning of day t (which is the information given by the sensor)

$\hat{a}_i^t$: expected daily filling rate of bin i at day t (in percentage of volume/day)

$\hat{F}_i^t$: expected fill level in percentage of volume at bin i at the end of day t ($S_i^t + \hat{a}_i^t$)

M: minimum fill level threshold for selecting the waste bins to be included in the dynamic subset L (in percentage of volume)

K: number of available vehicles

Decision variables

x_{ij}: binary variable indicating if edge (i, j) is visited, $(i, j \in I)$

y_{ij}: positive variable representing the flow between node i and node j, $(i, j \in I)$

g_i: binary variable indicating if waste bin i is visited, $(i \in I \setminus \{0, n+1\})$

Model

$$\max \mathrm{P} = R \sum_{i \in I \setminus \{0,n+1\}} \hat{F}_i^t E_i B g_i - 0.5 \left(C \sum_{i \in I} \sum_{j \in I, (j \neq i)} x_{ij} d_{ij} \right) \tag{1}$$

s.t.

$$\sum_{j \in I, (j \neq i)} (y_{ij} - y_{ji}) = 2\hat{F}_i^t E_i B g_i, \forall i \in I \setminus \{0, n+1\} \tag{2}$$

$$\sum_{i \in I \setminus \{0,n+1\}} y_{in+1} = \sum_{i \in I \setminus \{0,n+1\}} \hat{F}_i^t E_i B g_i \tag{3}$$

$$\sum_{j \in I \setminus \{0,n+1\}} y_{n+1j} = QK - \sum_{i \in I \setminus \{0,n+1\}} \hat{F}_i^t E_i B g_i \tag{4}$$

$$\sum_{i \in I \setminus \{0,n+1\}} y_{i0} = QK \tag{5}$$

$$\sum_{j \in I \setminus \{0,n+1\}} y_{0j} = 0 \tag{6}$$

$$\sum_{i \in I, (i \neq j)} x_{ij} = 2g_j, \forall j \in I \setminus \{0, n+1\} \tag{7}$$

$$y_{ij} + y_{ji} = Qx_{ij}, \forall i, j \in I, i \neq j \tag{8}$$

$$g_i = 1, \forall i \in I \setminus \{0, n+1\} : \hat{F}_i^{t+1} \geq 1 \tag{9}$$

$$x_{ij}, g_i \in \{0, 1\}, \forall i, j \in I, i \neq j \tag{10}$$

$$y_{ij} \in \Re^+, \forall i, j \in I, i \neq j \tag{11}$$

The objective function (1) considers the maximization of profit (P), defined as the difference between the revenues from selling the waste collected and the transportation cost (which is considered as a linear function of the distance travelled). Constraint (2) ensures that the difference between outflow and inflow at each bin is equal to twice the estimate weight, since this formulation considers the existence of two flows passing through each node. Constraint (3) ensures that the total inflow of the copy depot is equal to the total weight of waste in the visited bins, while constraint (4) ensures that the total outflow of the copy depot is equal to the residual capacity of the vehicle fleet. Constraint (5) guarantees that the total inflow of the real depot is equal to the capacity of the vehicle fleet and constraint (6) completes this by guaranteeing that the total outflow of real depot is equal to zero. The existence of two edges incident to each bin is ensured by constraint (7). Constraint (8) links variables x and y, guaranteeing that the sum of the flows for every edge (i, j) must be equal to the vehicles' capacity if the edge is traversed by the vehicle. Constraint (9) guarantees that a bin that is about to overflow, based on the information provided, will not remain in this state as a visit is imposed. The variables' domain is given in constraints (10) and (11).

4 Case Study

The developed methodology is applied to the same set of data as in [4]. The data pertains to a Portuguese recyclable waste collection company responsible for 14 municipalities. The company is characterized by a homogeneous fleet and a single depot. Despite performing three types of collection (undifferentiated, selective and organic), this work is focused on the selective collection of paper and cardboard.

Currently, this waste stream is collected through 26 static and pre-defined routes which are performed periodically, meaning that all waste bins present in a route are visited when this route is performed, without considering its fill-levels. However, the collection team is responsible for gathering information on the bin's fill-level, registering the visited waste bin as empty (0%), less than half (25%), half (50%), more than half (75%) and full (100%), at the moment of collection. From the route set, three were selected, specifically, routes 6, 11 and 13, as considered to be representative of the total operation. In total, they are composed of 226 bins: 68, 74 and 84 bins, respectively.

The time period to be analyzed was defined between January 3rd, 2013 ($t = 1$) and February 2nd, 2013 (30 days); during this period, route 6 was performed two times, route 11, three times and route 13, five times. Available data on the routes was analyzed and key performance indicators were attributed to each one. On Table 1, the total and average values are showed.Total values are the sum of the values for each route that was performed during a 30-day period, while average values are the total values divided by the number of performed routes, i.e. an average value per route.

From Table 1 it is concluded that the company collected more than 22 000 kg of waste and traveled more than 1 450 km in total, corresponding to a total profit of 636 euros. It can be also observed that a total of 81 bins were found empty and, on average, 8 empty waste bins were visited per route. In addition, roughly 66% of the collected bins presented a fill-level lower or equal to 50%, resulting in a 15.1 kg/km ratio.

Table 1 Current situation

KPI	Total	Average
Number of routes	10	1
Profit (euro)	636.3	63.6
Weight (t)	22.1	2.2
Distance (km)	1 468.4	146.8
Attended bins	778	78
Empty visited bins	81	8
Ratio (kg/km)	15.1	15.1
Vehicles used	10	1
Vehicles usage rate (%)	–	55.4

Most of the parameters considered for the application of the proposed solution method are the ones used in [4]. The first exception is parameter $\hat{a}_i^t$, which was calculated by [4] dividing the total filling levels registered for each waste bin by the number of days of the period. This parameter is here determined for the sub-period in-between collections, based on the fill-levels registered by the collection teams, through interpolation. Since after a collection the bins' fill-levels are zero, the expected daily filling rate is calculated as the fill-level of the next collection day divided by the number of days between the first and second collection in the subperiod. The $\hat{a}_i^t$ for the subperiod before the first collection considered was defined as the mean of all other $\hat{a}_i^t$ associated to that bin. The second exception is parameter S_i^1, that is here retrospectively estimated by subtracting the $\hat{a}_i^t$ of the subperiod before the first collection multiplied by the number of days between the first collection and the first day of the global period, to the fill-level registered in the first collection of that bin. As this last parameter depends on the values of $\hat{a}_i^t$, the values obtained in this work are different from those considered in [4].

As mentioned, in this work we also have the minimum fill-level threshold for selecting the waste bins to be included in the dynamic subset (M). Thus, a sensitivity analysis on this parameter M is performed, assuming values equal to 0%, 10%, 20%, 30%, 40% and 50% so as to understand how this value influences the decision and which value should be used. Note that, when $M = 0\%$, the approach proposed in the present work conceptually corresponds to the one applied by [4], where all bins are going to be considered within the route definition—226 bins.

5 Results and Discussion

Table 2 presents the results for the variation of parameter M, relevant for the formation of the L subset. Again, total values are the sum of the values for each route that was performed during a 30-day period, while average values represent the average value per route. By varying M, a total number of 6 solutions were studied. The mathematical model was implemented using GAMS 24.6.1 and solved with CPLEX Optimizer 12.6.3 on an Intel Xeon CPU X5680 @ 3.33 GHz, considering a limit of 16200 s as the maximum computational time allowed.

When comparing the solutions to the current situation (see Tables 1 and 2), one of the most obvious differences is the decrease in the number of routes performed (10 in the current situation and 6 to 11 for different M values). This would bring gains in both the economic and environmental dimensions. Economically, the operation would require less vehicles, used more efficiently; whereas environmentally the enhancement in the vehicle's usage rate implies a reduction in the CO_2, through the reduction of the number of routes.

Even though the model proposed by [4] is formulated to choose the number of vehicles for profit maximization, for comparability, the number of available vehicles in this work is fixed as one ($K = 1$). As a result, one vehicle is proved to be enough to assure that no bins overflow during the 30-day period, as the final planning requires

Table 2 Solutions for $M = 0\%$, $M = 10\%$, $M = 20\%$, $M = 30\%$, $M = 40\%$ and $M = 50\%$

KPI	$M = 0\%$		$M = 10\%$		$M = 20\%$	
	Total	Average	Total	Average	Total	Average
Number of routes	6	1	6	1	6	1
Profit (euro)	1 142.8	190.5	1 076.3	179.4	1 052.5	175.4
Weight (t)	22.6	3.8	22.3	3.7	21.4	3.6
Distance (km)	1 002.8	167.1	1 037.9	173.0	982.5	163.8
Attended bins	689	115	683	114	600	100
$\mid L \mid$	1 356	226	853	142	730	122
Ratio (kg/km)	22.5	22.5	21.4	21.4	21.8	21.8
Computational time (ks)	97.2	16.2	84.3	14.0	81.9	13.6
Vehicles used	6	1	6	1	6	1
Vehicles usage rate (%)	–	94.1	–	92.7	–	89.3
KPI	$M = 30\%$		$M = 40\%$		$M = 50\%$	
	Total	Average	Total	Average	Total	Average
Number of routes	6	1	6	1	11	1
Profit (euro)	988.8	164.8	960.3	160.0	732.9	66.6
Weight (t)	20.1	3.3	19.7	3.3	20.3	1.8
Distance (km)	918.0	153.0	910.5	151.7	1 197.1	108.8
Attended bins	452	75	400	67	376	34
$\mid L \mid$	550	92	442	74	467	42
Ratio (kg/km)	21.9	21.9	21.6	21.6	17.0	17.0
Computational time (ks)	58.8	9.8	57.0	9.5	34.1	3.1
Vehicles used	6	1	6	1	11	1
Vehicles usage rate (%)	–	83.6	–	82.1	–	46.2

only 6 routes for values of M between 0% and 40%. This, compared to the 10 routes that are performed in the current situation, means a reduction of 40% on the number of performed routes. The exception is for $M = 50\%$, representing an increase in route frequency. Also, the kg/km ratio is, for all studied M values, higher than the current situation (15.1 kg/km); with exception of $M = 50\%$ (17.0 kg/km), all the other values of M guarantees ratios kg/km higher than 21 kg/km (Fig. 2), which is a considerable improvement.

Regarding the economical dimension, a graph of the total profit as a function of M is displayed in Fig. 2. As expected, profit decreases as M increases, since less bins are feed to the routing model. The decision on whether to visit a bin depends on the contribution to the route profit, restrained by the capacity of the vehicle. The fewer the bins considered, the less likely is the existence of bins that would easily be integrated into the route generating profit, when revenue and cost are weighted. Nonetheless, when the profit per route or average profit are compared, the solution for $M = 50\%$ presents remarkably poor results, being 65% below that of solution $M = 0\%$ (Fig. 3). This can be justified by the fact that $M = 50\%$ leads to the definition of 11 routes,

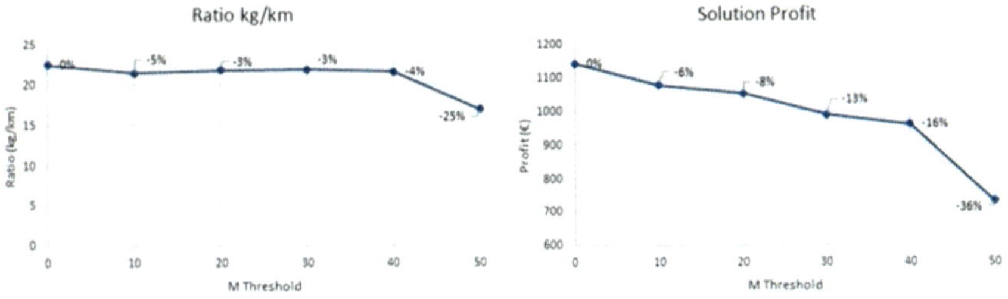

Fig. 2 Variation of ratio kg/km and of total profit, reporting to solution $M = 0\%$

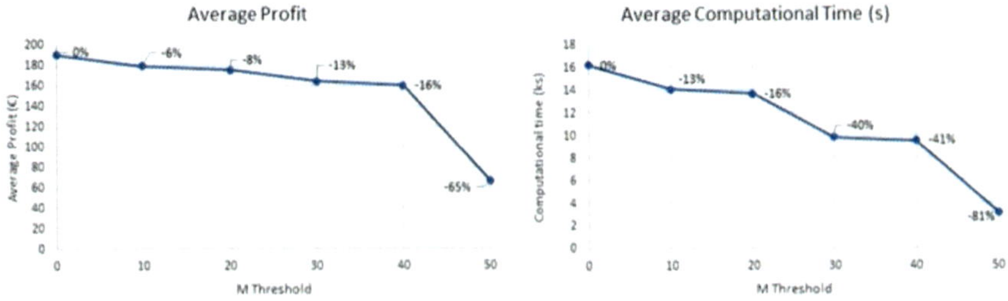

Fig. 3 Variation of average profit and of computational time, reporting to solution $M = 0\%$

whilst only 6 routes are defined for other M values. Compared to the other scenarios, $M = 50\%$ is the only scenario that performs more routes than the current situation (11 routes versus 10 routes).

Computationally speaking, the increase of M leads to, on average, less amount of time to obtain optimized routes (Fig. 3). Additionally, by increasing M, the number of solutions with gap = 0% tends also to increase (Fig. 4). The solution with the best computational performance is $M = 50\%$, with a reduction of required time higher than 80% and an enhancement on the percentage of solutions with gap = 0% of 81%. Nevertheless, when analyzing Fig. 4, it is possible to conclude that for $M = 50\%$, besides of the average profit (profit per route) that is much lower than that one for $M = 0\%$ because of the number of routes, as explained before, the same also happens for the vehicle usage rate. This because less bins are visited per route since the dynamic set L to be considered for the routing model becomes smaller as M increases. In this sense, for $M = 50\%$, an additional burden to the environment is indicated as more collection routes are defined to be performed, collecting however lower amount of waste bins.

Thus, the main conclusion regarding the optimal parameter M is that it should not surpass the value of 40%. Solutions for $M = 30\%$ and 40% have similar performances in all KPI, proving that the ideal value should be between those two values.

When comparing the results obtained by [4] for scenario 3A with the ones obtained in this study it can be concluded that our proposed approach performs better. However, to allow a fair comparison scenario 3A explored by [4] is tested using the same values for parameters S_i^1 and $\hat{a}_i^l$ that were considered in this work. As an example,

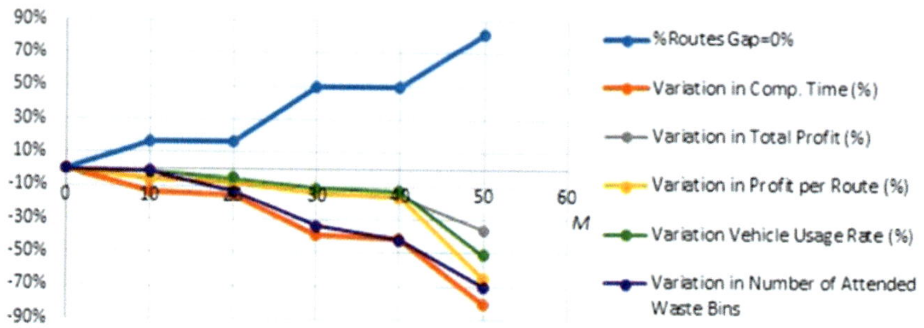

Fig. 4 KPIs' summary, as a variation reporting to the solution of $M = 0\%$

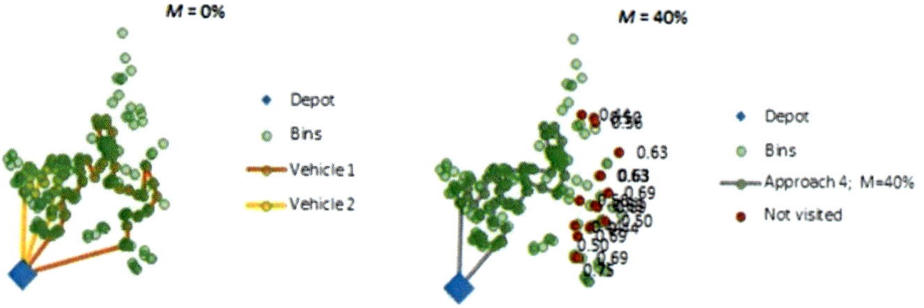

KPI	M=0% Day 1	M=40% Day 1
Profit (€)	261.0	230.5
Attended bins	138	71
|L|	226	88
Ratio (kg/km)	22.5	28.4
Gap (%)	8.0	0.0
Computational time (s)	14400.0	4415.6
Vehicles used	2	1

Fig. 5 Day 1 routes comparison between scenario 3A and our proposed approach

a comparison is done for the routes performed at day 1 when applying the solution approach explored by [4] and the approach proposed in this present work, using $M = 40\%$ (Fig. 5). In this case, to allow a fair comparison, the number of available vehicles is defined to be equal to two, as in [4] (Eqs. 4 and 5 are in this case adapted to be inequations, allowing the model to choose the number of vehicles to use). It is possible to see that, for the same day, our proposed approach decides to collect less waste bins (71 against 138 from scenario 3A) using just one vehicle. Besides this, and differently from scenario 3A where all 226 waste bins are feed to the routing model, when applying our proposed model using $M = 40\%$ only 88 waste bins are feed to the model and, from this dynamic waste bins set, the model decided not to collect 17 waste bins (highlighted in red in Fig. 5) because they were not worth to visit. Despite less profit is obtained at day 1 when applying our approach, a higher kg/km ratio is obtained even if less waste bins are collected. This represents both economic and environmental gains: less vehicles are required and the vehicles are used more efficiently, implying on a CO_2 reduction.

6 Conclusions

In the present work, an optimization based-heuristic is proposed to solve the SWCRP to optimality in reasonable times when an operational level is at stake. The proposed approach decomposes the SWCRP by first reducing the number of bins to be considered for collection; and then, by applying the selected number of bins to a VRPP model choosing to visit only those waste bins that are worth to be collected when considering their filling levels and location.

When comparing the results obtained using real data from a case study, it is possible to conclude that local optimal solutions were obtained in a reasonable time and that the method developed brings gains in both the economic and environmental dimensions, when compared to the current situation of the real case. Economically, the operation generally requires less vehicles, used more efficiently, whereas environmentally the enhancement in the vehicle's usage rate implies a reduction in the CO_2, through the reduction of the number of routes. Besides this, an interesting conclusion is obtained on the value of M, when a fixed number of vehicles is considered. After performing a sensitivity analysis was proved that a value between 30% and 40% results in a good compromise between computational performance and profit. However, as if the number of vehicles is allowed to vary, profit could be impacted and consequently there might be compromise regarding overall profit and profit per route. Thus, as future work a worth path to explore is to study the value of M considering the number of vehicles as a variable.

References

1. Korhonen, J., Honkasalo, A., Seppälä, J.: Circular economy: the concept and its limitations. Ecol. Econ. **143**, 37–46 (2018)
2. MacArthur, E., Zumwinkel, K., Stuchtey, M.R.: Growth within: a circular economy vision for a competitive Europe. Ellen MacArthur Foundation (2015)
3. Gonçalves, D.S.: Tecnologias de informação e comunicação para otimização da recolha de resíduos recicláveis Master's thesis. Lisboa: ISCTE-IUL (2014)
4. Ramos, T.R.P., de Morais, C.S., Barbosa-Póvoa, A.P.: The smart waste collection routing problem: alternative operational management approaches. Expert Syst. Appl. **103**, 146–158 (2018)
5. Johansson, O.M.: The effect of dynamic scheduling and routing in a solid waste management system. Waste Manage. **26**, 875–885 (2006)

6. Hannan, M.A., Mamun, M.A.A., Hussain, A., Basri, H., Begun, R.A.: A review on technologies and their usage in solid waste monitoring and management systems: issues and challenges. Waste Manage. **43**, 509–523 (2015)
7. Melaré, A.V.S., González, S.M., Faceli, K., Casadei, V.: Technologies and decision support systems to aid solid-waste management: a systematic review. Waste Manage. **59**, 567–584 (2017)
8. Dondo, R., Cerdá, J.: A cluster-based optimization approach for the multi-depot heterogeneous fleet vehicle routing problem with time windows. Eur. J. Oper. Res. **176**, 1478–1507 (2007)
9. Fisher, M.L., Jaikumar, R.: A generalized assignment heuristic for vehicle routing. Networks **11**(2), 109–124 (1981)
10. Hu, X., Huang, M.: An intelligent solution system for a vehicle routing problem in urban distribution. Int. J. Innov. Comput., Inf. Control. **3**(1), 189–198 (2006)
11. Beasley, J.E.: Route first - cluster second methods for vehicle routing. OMEGA **11**(4), 403–408 (1983)
12. Baldacci, R., Hadjiconstantinou, E., Mingozzi, A.: An exact algorithm for the capacitated vehicle routing problem based on a two-commodity network flow formulation. Oper. Res. **52**, 723–738 (2004)

A Multi-objective and Multi-period Model to the Design and Operation of a Hydrogen Supply Chain: An Applied Case in Portugal

Diego Câmara, Tânia Pinto-Varela, and Ana Paula Barbosa-Póvoa

Abstract Hydrogen has been listed as one of the main energy alternatives to integrate the world energy matrix, with great relevance in the automotive sector. This integration aims to contribute for the reduction of global potential warming through global supply chains (SC) that support industrial processes using renewable energy sources. Hydrogen is an option with a high potential to mitigate environmental impacts caused by the current fossil fuels. Although it has benefits in environmental terms, there are no current infrastructures and does the design of such network requires high capital investment, being this the main barrier to the hydrogen economy development. It is then crucial to define a future hydrogen SC in an optimized way. Following this need, we propose a multi-objective, multi-period mathematical formulation for the design and planning of a hydrogen SC. A Mixed Integer Linear Programming (MILP) model is developed to determine the Hydrogen SC planning and operational decisions. The formulation minimizes cost and environmental impacts of the network. A case study in Portugal was explored.

Keywords Hydrogen supply chain · Multi-objective optimization · Renewable energy source

1 Introduction

In the last decade, the need to replace fossil fuels with environmentally friendly alternatives has been much discussed. Scientific advances in technologies associated with new types of clean fuels are increasing in terms of production, distribution or storage. Hydrogen is one of these fuel alternatives for transportation systems with significant potential to contribute to a more sustainable world energy matrix. Although hydrogen has advantages in energetic and environmental terms, there are some barriers that affect the development of the so-called hydrogen economy. The lack of avail-

D. Câmara · T. Pinto-Varela (✉) · A. P. Barbosa-Póvoa
Centro de Estudos de Gestão-IST, Instituto Superior Técnico, Universidade de Lisboa, Av. Rovisco Pais, 1049-001 Lisboa, Portugal
e-mail: tania.pinto.varela@tecnico.ulisboa.pt

S. Relvas et al. (eds.), *Operational Research*, Springer Proceedings in Mathematics & Statistics 374, https://doi.org/10.1007/978-3-030-85476-8_2

able infrastructure and the existence of uncertainties associated with the strategic planning of the hydrogen network have been the main identified barriers [1]. In an attempt to mitigate the economic barrier different approaches have been formulated and discussed in the literature.

The formulation developed by Almansoori and Shah [2] called "Snapshot" is regarded as the base work in design and planning of hydrogen SC. In their subsequent studies, the prior formulation was extended considering: primary energy sources availability (PES); its logistics, and variation of hydrogen demand [3]. Fueling stations number and hydrogen local distribution between different regions was added to the formulation in [4]. Since 2008, single-objective optimization approaches for the design and planning of the hydrogen SC have been developed with the cost as objective [4–7]. More recently and accounting for the recent environmental concerns, new formulations have emerged exploring multi-objective models in which compromise solutions are targeted considering not only the financial aspect but also the potential environmental damage. Guille et al. [8] formulated a bi-objective problem, which simultaneously accounts for cost and environmental impact minimization. A bi-level algorithm is also introduced to expedite the search of Pareto solutions through a decomposition algorithm. Ogumerem et al. [9] considered the maximization of the net present value (NPV) and the minimization of the greenhouse gas (GHG) emissions. A Pareto front was developed and the extreme points on the curve were used to test two scenarios. One of those, if the oxygen co-produced would be sold generating revenues, and the other one, if discarding oxygen would be the most economical strategy. Sabio et al. [10] proposed a framework for optimizing hydrogen SC according to several environmental indicators, considering also economic performance. Kim et al. [6] formulated a MILP model to identify the optimal supply chain configurations considering cost efficiency, and safety. The study was applied to an expected future hydrogen infrastructure in Korea. Sabio et al. [11] introduced a support tool to strategic planning of hydrogen SC with uncertainty. Besides considering cost, a risk metric was considered as an additional objective function. Almaraz et al. in [12] solved a tri-objective optimization problem, considering cost, environmental impact and safety objective functions. The results were compared with the study case proposed before by Almansoori and Shah. Almaraz et al. in [13] extended their tri-objective work in a multi-period purpose and build a Pareto front through lexicographic optimization. In this case, the multi-period formulation turned out to be not feasible due to the problem size and the use of binary variables. Finally, Almaraz et al. in [14] extended their multi-objective approach, and a comparison was performed between different geographic scales, from regional to national. A sequential optimisation to build the SC expansion over time was considered.

From our knowledge, a gap still remains to fulfil, this related to the use of a multi-objective approach considering economic and environmental goals within a multi-period context.

In this perspective, to offer a compromise solution with the current environmental awareness, network deployment and operation cost, this work proposes a multi-objective, multi-period formulation. Global warming potential (GWP) was taken as the environmental impact measure, which is the overall effect related to emissions of greenhouse gases (CO_2-equiv).

2 Problem Characterization

According to the literature, hydrogen SC is characterized by several echelons, see Fig. 1, with a high level of interaction between them.

Hydrogen's production is supported by different energy sources, in this work considered as renewables energies (solar power, wind power, hydroelectric and biomass). Based on this energy set, an appropriate hydrogen production technology should be selected. Hydrogen produced is then only distributed between different regions, meaning that no distance between production plants and local storage facilities is considered. A multimodal transportation mode is considered. Finally, according to its physical form, gaseous or liquid, the hydrogen produced is stored in storage facilities.

Hydrogen demand has a deterministic estimated profile calculated, which is assumed to be demand driven, in other words, demand must be satisfied.

In order to serve as a support tool for the decision maker, the total design and operation cost of the hydrogen network, and the measurement of environmental impacts are comprised. A multi-objective approach provides a trade-off solution considering a multi-period formulation. The problem can be described as follows:

Given:

- A set of locations to install hydrogen production units, and storage facilities;
- Type of production and storage facilities (electrolysis, biomass gasification);
- Maximum and minimum capacities for flows, production, and storage;
- Global warming potential for each transportation mode;
- Regional delivery distances;
- PES availability;
- Operating and capital costs;
- Customer's demand.

Fig. 1 Typical structure of hydrogen supply chain. Adapted from [15]

Determine:

- The number, location, capacity, type of technologies for hydrogen production plants and storage facilities;
- The network planning with all flows, rates of hydrogen and PES consumption, production rates and average inventory of materials.

So as to:

- Minimize the cost of the supply chain and the GWP.

3 Methodology

In this section, the developed methodology is presented: model approach and solution approach.

3.1 *Model Approach*

A multi-period, bi-objective formulation is developed. The mathematical formulation is composed by a set of constraints: mass balance, production, transportation and storage constraints, PES availability and demand requirements. Mass balance: determine the production rate, relating flows of hydrogen in each region and the respective demand. Production and storage: define production capacity limits, relating the number of production plants and storage facilities with their capacities. Transportation: handle hydrogen flows between different regions to satisfy the demand, within capacity limits. This work aims to minimize the average daily total cost (TDC), as well as, the environmental impact of the network, Eqs. 1 and 2, respectively.

$$\mathbf{min} TDC = \frac{1}{NTP}\left(\frac{FCC + TCC}{B * CCF} + FOC + TOC + ESC\right) \tag{1}$$

$$\mathbf{min} TotalGWP = \frac{1}{NTP}(PGWP + TGWP + SGWP) \tag{2}$$

Equation 1, characterizes the total sum related to the supply chain costs, involving capital costs associated with facilities and transportation, FCC and TCC, respectively. Moreover, operating costs that involve facilities, transportation and energy sources are considered: FOC, TOC and ESC, respectively. Once the model comprises operational, besides strategic decisions, the cost is presented in diary value. Thus, capital costs, FCC and TCC are transformed into daily values through consideration of a capital charge factor (in years), CCF, related to the investment payback, and the number of SC operational days in a year, B.

Equation 2, minimizes the environmental impacts totalized by the sum of GWP related to production, transportation and storage: PGWP, TGWP, SGWP, respectively.

For reasons of brevity, components of each objective function are not detailed. The number of time periods, NTP, is considered in the objective functions to reach an average value for each time period.

3.2 Solution Approach

Among the available methods to deal with multi-objective problems, the ϵ-constraint method was chosen given the simplicity and applicability of its implementation. It consists in optimizing the cost objective function using the environmental impact function as constraint, and then by varying the constraint bounds, the points of Pareto front were obtained. In addition, to guarantee the efficiency of the obtained solutions avoiding weak (dominated) solutions, a lexicographic optimization was performed, defining an efficient Pareto front [16].

4 Case Study

A case study for the design and planning of a hydrogen supply chain was explored over Portugal. A district segmentation of Portugal was utilized, with 18 districts. A realistic path between districts and the existing main roads and truck lines is assumed. A deterministic PES availability is considered according to the current data of Portugal. Variability over time is not taken into account.

As showed in Fig. 2, four renewable energy sources are considered: wind power, solar power, hydroelectric and biomass. Two types of production technologies are explored: electrolyzes and biomass gasification. Hydrogen is produced in liquefied form. A tanker truck and a railway tanker are used as transportation modes. Some regions in the north of Portugal, namely, regions 3, 4, 7, 8 shown in Fig. 4, are assumed not to have railway infrastructure. Cryogenic storage is assumed (Table 2).

To hydrogen demand estimation the current number of private and light vehicles in Portugal, the average distance travelled, and hydrogen fuel economy (kg H_2/km) was used. The deterministic hydrogen demand is based on a hydrogen penetration factor in each time period. Three periods were considered, with 5 years each. The hydrogen penetration factor is 5%, 25% and 50%, for the first, second and third period, respectively, for 15 years of network planning. CCF is assumed to be 5 years.

The model was solved through GAMS 25.1.1, using CPLEX 12.0, in a two Intel Xeon X5680, 3.33 GHz computer with 24 GB RAM. Computational statistics are summarized in Table 4.

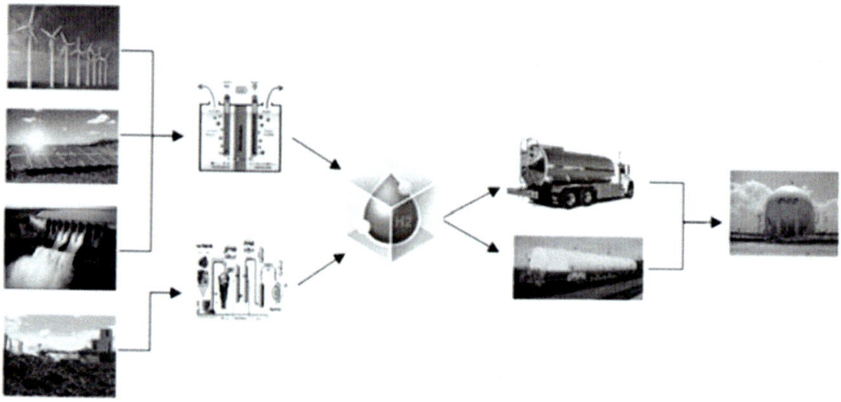

Fig. 2 Hydrogen supply chain characterization applied to the case study

Table 1 Production Plants capacities

Production technology/Sizes	Biomass		Electrolysis	
	M	L	S	M
Minimum capacity (t/d)	10	200	0.3	10
Maximum capacity (t/d)	150	960	9.5	150

Table 2 Storage facilities capacities

Sizes of storage	S	M	L
Minimum storage capacity (t/d)	0.5	10	200
Maximum storage capacity (t/d)	9.5	150	540

5 Results and Discussion

The model results are analysed considering average values for TDC and Total GWP, as represented in Fig. 3 and Table 3. The first and most important analysis to be done is on the trade-off between the different solutions selected in terms of network cost and GWP, which explains the usage of this tool to support strategic decision.

The simplified Pareto front is represented in Fig. 3. From this Pareto front approximation, three solutions were identified for further analysis (point 1, 4 and 6). Each point represents a multi-period solution, which embraces all planning information and strategic decisions of the SC expansion for three time periods. Points 1 and 6 characterize respectively solutions with maximum and minimum values of total daily cost. Point 4 depicts the evolution of the SC topology between those two multi-period solutions selected. The SC network characterization in terms of topologies, costs and GWP details are shown in Table 3.

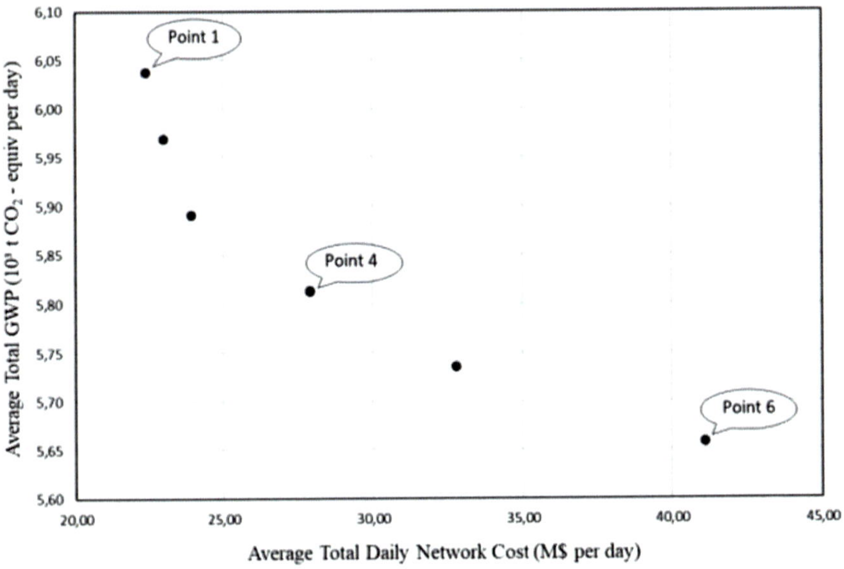

Fig. 3 Pareto front approximation

From point 1 to point 6, a cost increase of almost 84% is observed while a 6% of GWP reduction was reached. Between points 1 and 4, there is an increase of approximately 25% in the total daily cost, to achieve a reduction of about 4% in GWP. From these results, it is verified that the impact of cost is very high when compared to the reduction achieved in the average GWP.

The topology evolution for the multi-period solution point 1 is detailed in Fig. 4. This network has different types of production plants, requiring different biomass gasification and electrolysis production technologies, to offer a solution with a compromise to the most possible lowest cost. In contrast, point 6 topology, presents only one production plant type, consequently one production technology, the electrolysis. In this case, the biomass is not an option as energy source. Despite its lower price, biomass gasification is a less environmental friendly production technology when compared to electrolysis.

Related to the network decisions it is possible to observe that the number of production plants increases from 9 to 71 from point 1 to 6, as shown in Table 3. This result is justified by the decrease in facilities number using biomass as an energy source, which it is also associated with technologies capacities. As seen in Table 1, production plants with electrolysis technology have lower productive capacities.

The number of storage facilities remains the same for all solutions points, from 1 to 6, since the demand variability is unchanged between the different point solutions.

It still observed that long transportation links are established between regions as this option is cheaper than open new production plants.

Table 3 Comparison of multiperiod solutions topologies from the Pareto front

Point in Pareto Frontier	1	4	6
Number of production facilities	9	15	68
Electrolysis Technology	5	14	68
Biomass gasification	4	1	0
Number of storage facilities	67	67	67
Number of transportation units	295	143	0
Maintenance cost ($ per day)	5497	2153	0
Fuel cost ($ per day)	12044	3985	0
Labour cost ($ per day)	24377	11668	0
General cost ($ per day)	2053	990	0
Capital cost			
Plants and storage facilities (M$)	111,560	136,200	201,480
Transportation modes (M$)	147.50	71.50	0.00
Operating cost			
Plants and storage facilities (M$ per day)	5.85	9.05	12.91
Transportation modes ($ per day)	43972	18977	0
Average Total Daily Network Cost (M$ per day)	22.37	27.91	41.10
Average GWP Production facilities (10^3 t CO_2—equiv per day)	2.02	1.82	1.68
Average GWP Storage facilities (10^3 t CO_2—equiv per day)	3.98	3.98	3.98
Average GWP Transportation modes (10^3 t CO_2—equiv per day)	0.034	0.013	0.000
Average Total GWP (10^3 t CO_2–equiv per day)	6.04	5.81	5.66

Table 4 Computational statistics for the multi-objective optimization

Continuous variables	Integer variables	Constraints	CPU (d)	GAP (%)
9683	3025	12653	7, 5	0.0

The results suggest decentralized SC topologies for lower values of GWP (point 6), with smaller capacities and where the demand for hydrogen is supplied locally by production plants. However, the network cost minimization suggests a centralized topology with higher capacities, as in point 1. Also, it is verified that the number of transportation units, is inversely associated with the number of plants.

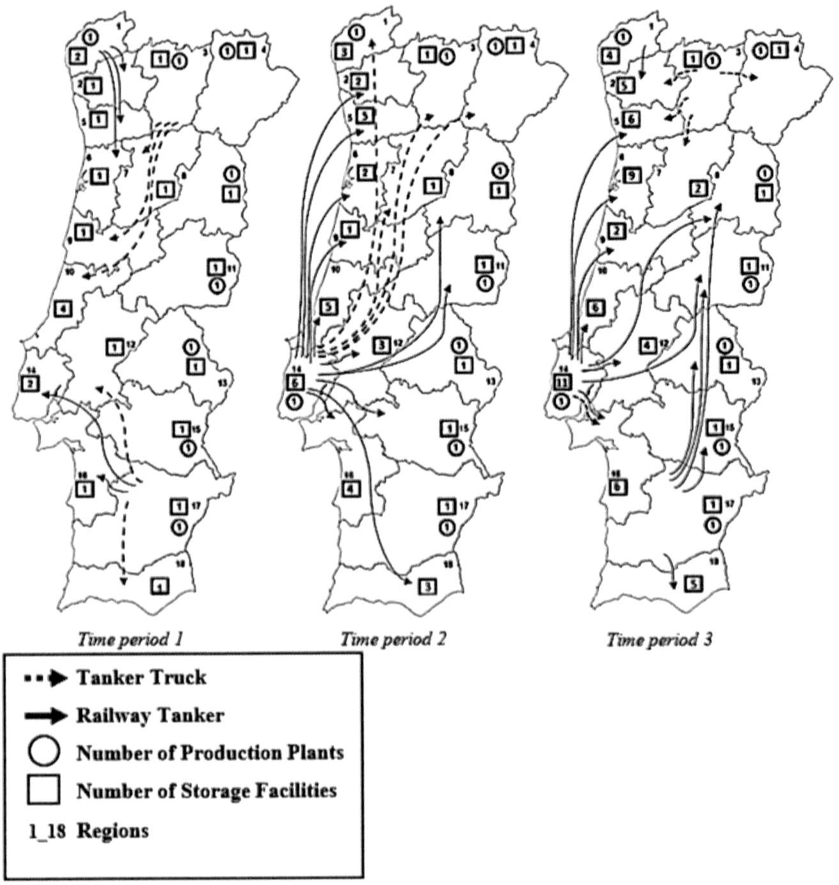

Fig. 4 Topologies of solution point 1 over time

6 Conclusions

In this work, a MILP model to support the decision maker in the design and planning of hydrogen supply chains is developed. The model considers the trade-off between economic and environmental objectives. A case study exploring a future hydrogen supply chain in Portugal was studied.

The results show that a high increment in cost is required when compared to the decrease in environmental impact. In terms of topology analysis, it is found that when a cost objective is at stake a centralized network is obtained, however when the environmental objective is the main goal, a decentralized supply chain is obtained.

As future work, new features in the formulation are intended to be added, for instance, politic incentives to a hydrogen economy, thus analyse their impact in the SC network. It is also possible and expected to integrate a multicriteria analysis method to assist the model. Also, uncertainty sources in different aspects of the chain should be studied, as well, the presence of risk should be considered so as to increase the model accuracy to deal with forecasts associated to design of hydrogen SC.

References

1. Nunes, P., Oliveira, F., Hamacher, S., Almansoori, A.: Design of a hydrogen supply chain with uncertainty. Int. J. Hydrogen Energy **40**, 16408–16418 (2015)
2. Almansoori, A., Shah, N.: Design and operation of a future hydrogen supply chain Snapshot Model. Chem. Eng. Res. Des. **84**(6), 423–438 (2006)
3. Almansoori, A., Shah, N.: Design and operation of a future hydrogen supply chain: multi-period model. Int. J. Hydrogen Energy **34**(19), 7883–7897 (2009)
4. Almansoori, A., Shah, N.: Design and operation of a stochastic hydrogen supply chain network under demand uncertainty. Int. J. Hydrogen Energy **37**(5), 3965–3977 (2011)
5. Kim, M., Kim, J.: Optimization model for the design and analysis of an integrated renewable hydrogen supply (IRHS) system: application to Korea's hydrogen economy. Int. J. Hydrogen Energy **41**(38), 16613–16626 (2016)
6. Kim, J., Lee, Y., Moon, I.: Optimization of a hydrogen supply chain under demand uncertainty. Int. J. Hydrogen Energy **33**(18), 4715–4729 (2008)
7. Samsatli, S., Staffell, I., Samsatli, N.J.: Optimal design and operation of integrated wind-hydrogen-electricity networks for decarbonising the domestic transport sector in Great Britain. Int. J. Hydrogen Energy **41**(1), 447–475 (2015)
8. Guille, G., Mele, F.D., Grossmann, I.E.: A bi-criterion optimization approach for the design and planning of hydrogen supply chains for vehicle use. Am. Inst. Chem. Eng. J. **56**(3), 650–667 (2010)
9. Ogumerem, G.S., Kim, C., Kesisoglou, I., Diangelakis, N.A., Pistikopoulos, E.N.: A multi-objective optimization for the design and operation of a hydrogen network for transportation fuel. Chem. Eng. Res. Des., 1–14 (2017)
10. Sabio, N., Kostin, A., Guille, G.: Holistic minimization of the life cycle environmental impact of hydrogen infrastructures using multi-objective optimization and principal component analysis. Int. J. Hydrogen Energy **37**, 5385–5405 (2011)
11. Sabio, N., Gadalla, M., Guille, G.: Strategic planning with risk control of hydrogen supply chains for vehicle use under uncertainty in operating costs: a case study of Spain. Int. J. Hydrogen Energy **35**, 6836–6852 (2010)
12. Almaraz, S.D.-L., Azzaro-Pantel, C., Montastruc, L., Pibouleau, L., Senties, O.B.: Assessment of mono and multi-objective optimization to design a hydrogen supply chain. Int. J. Hydrogen Energy **38**(33), 14121–14145 (2013)
13. Almaraz, S.D.-L., Azzaro-Pantel, C., Montastruc, L., Domenech, S.: Hydrogen supply chain optimization for deployment scenarios in the Midi-Pyrénées region, France. Int. J. Hydrogen Energy **39**(23), 11831–11845 (2014)
14. Almaraz, S.D.-L., Azzaro-pantel, C., Montastruc, L., Boix, M.: Deployment of a hydrogen supply chain by multi-objective/multi-period optimisation at regional and national scales. Chem. Eng. Res. Des. **104**, 11–31 (2015)
15. Li, L., Manier, H., Manier, M.A.: Hydrogen supply chain network design: an optimization-oriented review. Renew. Sustain. Energy Rev. **103**(June 2018), 342–360 (2019)
16. Mavrotas, G.: Effective implementation of the e -constraint method in multi-objective mathematical programming problems. Appl. Math. Comput. **213**(2), 455–465 (2009)
17. Kim, J., Moon, I.: Strategic design of hydrogen infrastructure considering cost and safety using multiobjective optimization. Int. J. Hydrogen Energy **33**(21), 5887–5896 (2008)

The $\varepsilon-$Constrained Method to Solve a Bi-Objective Problem of Sustainable Cultivation

Angelo Aliano Filho and Helenice Florentino Silva

Abstract This study presents a nonlinear bi-objective 0-1 optimization model for sustainable cultivation and proposes an exact method to solve it. In this formulation, among a set of cultures, a predefined number of cultivable plots and planning horizon, it is intended to decide which crops, periods and plots should be cultivated. Two conflicting objectives are considered: (i) minimize the proliferation of pests and (ii) maximize the profit of the planting schedule in all planning horizon. The mathematical formulation was solved by the classical $\varepsilon-$constrained method. We linearized the original model and obtained an alternative linear version of our problem. Then, we compare the performance of $\varepsilon-$constrained method in this two formulation to determine some Pareto optimal solutions in 27 instances generated by a semi-random procedure of real dimension. The experiments showed that mathematical models along with the proposed method may be powerful tools in the complex decision-making in this field.

Keywords Multi-objective optimization · $\varepsilon-$Constrained method · Sustainability

1 Introduction

In the current context of sustainability and a cultivation practice in agriculture that minimizes the environmental degradation, alternative ways that avoid the intensive use of chemical products in combating pests and that increase the use of the soil are being strongly studied. These measures, can stop an environmental degradation in the planet and make agriculture products better.

In this sense, one of the central focuses in the crop production discussed lately is the use of measures that aim a sustainable and ecological planning, considering the

A. A. Filho (✉)
Federal Technological University of Paraná, Apucarana, Paraná, Brazil
e-mail: angeloaliano@utfpr.edu.br

H. F. Silva
Institute of Bioestatistics of Botucatu, State University of São Paulo, São Paulo, Brazil
e-mail: helenice@ibb.unesp.br

S. Relvas et al. (eds.), *Operational Research*, Springer Proceedings in Mathematics & Statistics 374, https://doi.org/10.1007/978-3-030-85476-8_3

Fig. 1 Crop rotation cultivation Available in: https://www.google.com/search?q=crop+rotation&rlz=1C1PRFE_enBR771BR771&source=lnms&tbm=isch&sa=X&ved=0ahUKEwityda0y__hAhVGH7kGHSZoCNsQ_AUIDigB&biw=1280&bih=689#imgrc=JyfqP2zMa_LLJM: access in May 3rd, 2019

environmental degradation that has occurred in recent years. For this reason, the planning of agriculture activities, among them the *crop rotation*, has gained prominence in the studies for sustainable cultivation, since it is one of the means of cultivation whose practical principles enable an ecological and productive agriculture. This practice, once conducted by the rural farmers, brings many benefits, since the control of pests, pathogens and weeds is performed biologically, decreasing the action of pesticides harmful to the environment and bringing measures of soil recovery, making it always fertile, as stated the work in [8].

The Fig. 1 illustrates a area with crop rotation system production.

Note that, in this area, there is a variety of different crops being cultivated, thus generating a scenario quite heterogeneous. As the crops are harvested, others are planted hence exploring the soil differently. According the studies reported in [3] and [9] these diversity also increase the soil productivity.

Techniques of operation research may help the decision makers in this complicated areas. Many times, the problems generated in this field are not trivial and involves decisions which can affect ecosystems, already accomplished planning, and bring about change in the supply chain at various levels. Such problems can have a mathematical representation and deterministic methods to optimize one or more functions can be used.

In this perspective, we can find in the literature some previous work in this direction. For example, the paper in [7] proposed an 0-1 optimization model to find a schedule of crop rotation to each plot in a crop area, maximizing the land occupation time. In the formulation proposed, are included planting constraints for adjacent lots, for crop sequences, green manuring and fallow periods. The work [6] also proposed a scheme of crop rotation, with an addiction of constraint of meeting the demand, respecting the same constraints of the previous study. Due to the large number of variables and constraints of the problems, they used the column generation method to solve it.

Other related study with operational research in sustainable development was reported in [5]. The authors proposed an integer programming formulation for crop rotation, considering both decisions on plot sizes and schedules simultaneously. A branch-price-and-cut algorithm was developed and extensive computational experiments over a set of instances based on real-life data were done.

The study [1] also addresses crop rotations, with the objective of maximizing the profit of rotation, incorporating the same ecological constraints of the model presented in [7]. The authors proposed several metaheuristics to solve the problem, as well as a few greedy constructive heuristics.

Approaches with multiple objectives in this field also has been introduced, and make the models more realistic. Hardly the problems of this nature involves to optimize a single objective-function. The work in [2] presents a review of the multicriteria analysis applied to agricultural resource management. Methods for selecting multiattribute discrete alternatives and for solving multi-objective planning problems are revised. Criteria used for modeling agricultural systems and to identify the difficulties for practitioners in applying the methodology are developed. The paper in [2] propose multi-objective linear programming models that consider a wide range of farming situations, which allows optimization of profit or environmental outcome, or both. The objective is to identify the best cropping and machinery options which are both profitable and result in improvements to the environment.

Differently of the previous works, the present paper consider economic and environmental objectives to be optimized concomitantly. In this sense, a nonlinear (more specifically quadratic) bi-objective 0-1 optimization model inspired in the studies in [1], [2] and [5] aims to minimize (i) the proliferation of pests among the planted crops (called z_1 function) and (ii) maximize the profit of the planting schedule (called z_2 function) in a certain planning horizon. Constraints of non-overlapping planting, non-consecutive planting of varieties of the same family and of the planning time equal in all plots are also considered.

To determine some Pareto optimal solutions to this problem, we applied the classical $\varepsilon-$constrained method, i.e., we consider the z_1 function as objective for the scalar problem and treated the z_2 function limited by a lower bound as an additional constraint. Varying this lower bound, we can find different Pareto optimal solution for this problem and offer a broad and varied scenario for the decision maker.

In summary, this work brings the following innovative aspects:

- adaptation of a 0-1 bi-objective mathematical programming model for sustainable cultivation;
- a comparative computational study between the quadratic and linear formulations of the adapted model
- application of an exact technique for solve the mono-objective problems.

The remainder of this article is organized as follows. Section 2 presents the mathematical modeling of the problem to be solved, the linearization technique employed to obtain Pareto optimal solutions. Section 3 describes the implementation and adaptation of the $\varepsilon-$constrained method in this problem. Section 4 presents the computational results obtained with 27 instances generated to validate the mathematical model and the methods developed. Finally, Sect. 5 describes the conclusions added of new study perspectives.

2 Mathematical Modeling

In this problem, we aim to minimize the infestation of pests among the culture in the stipulated planning horizon in all the planting area and, at the same time, maximize the profit of the planting schedule. Ecological constraints of alternate planting of distinct crops consecutively are considered, in addition to the non-overlapping planting and finally constraints that force all the plots to have the same duration cycle.

As discussed in previous paragraphs, this mathematical model was inspired in the study in [1] (which is mono-objective and aims to maximize the period of occupation of the cultivated area) and in [7] (which only aims to maximize the profit of the schedule). The first objective-function of this formulation and one constraint set are something peculiar of this study.

To model this problem, we defined the following indexes and parameters:

Indexes:

- i and $\bar{i}$: related to the crops;
- j and $\bar{j}$: related to the lots;
- t: related to the planning horizon periods;
- p: related to the botanical family of plants;
- r: intrinsically related to summation of variables in the constraints of the model.

Parameters:

- N: number of crops available for planting;
- K: number of available lots;
- T: duration of the planning horizon;
- NF: number of botanical family of plants;
- F_p: set of crops belonging to the family p, $p = 1, \ldots, N_f$;
- C_i: duration of culture life cycle i;
- l_i: profitability of culture i per hectare;
- $area_j$: area of plot j in hectare;
- $P_{i\bar{i}}$: in this study, we propose a measure of probability of pests infestations among the cultures i and $\bar{i}$. The infestation tend to have a higher probability between similar cultures rather than a lower probability when the crops are botanically different. Thus, this probability of infestation is given by:

$$P_{i\bar{i}} = \begin{cases} 0.9, & \text{if } i \text{ and } \bar{i} \text{ are from same families,} \\ 0.5, & \text{if } i \text{ and } \bar{i} \text{ are from families of degree between 2 and 3,} \\ 0.1, & \text{if } i \text{ and } \bar{i} \text{ are families with degree greater than 4,} \end{cases}$$

where the degree between two cultures i and $\bar{i}$ ($i \neq \bar{i}$) is defined as the difference (in module) between the families to which they belong.

- $V_{j\bar{j}}$: is a measure, proposed in this study, to determine the probability of pest infestation between the plot j and $\bar{j}$, $j \neq \bar{j}$, and given by

$$V_{j\bar{j}} = \frac{1}{1 + d_{j\bar{j}}},$$

in which $d_{j\bar{j}}$ d is the smallest distance between them. Note that, if $d_{j\bar{j}} \to 0$, then $V_{j\bar{j}} \to 1$; if $d_{j\bar{j}} \to \infty$, then $V_{j\bar{j}} \to 0$. This means that lots closer together tend to have a higher infestation.

The decision-making variables of this problem are defined as follows:

$$x_{ijt} = \begin{cases} 1, & \text{if crop } i \text{ is planted in plot } j \text{ in the period } t, \\ 0, & \text{otherwise,} \end{cases}$$

for all $i = 1, \ldots, N$, $j = 1, \ldots, K$ and $t = 1, \ldots, T$.

The nonlinear formulation of this problem is presented in (1)–(6).

$$\text{minimize } z_1 = \sum_{i=1}^{N} \sum_{\bar{i}=1}^{N} \sum_{j=1}^{K-1} \sum_{\bar{j}=j+1}^{K} \sum_{t=1}^{T} P_{i\bar{i}} \cdot V_{j\bar{j}} \cdot x_{ijt} \cdot x_{\bar{i}\bar{j}t} \tag{1}$$

$$\text{maximize } z_2 = \sum_{i=1}^{N} \sum_{j=1}^{K} \sum_{t=1}^{T} l_i \cdot area_j \cdot x_{ijt} \tag{2}$$

$$\text{subject to } \sum_{i=1}^{N} \sum_{r=0}^{C_i-1} x_{ij(t-r)} \leq 1, \quad j = 1, \ldots, K, \quad t = 1, \ldots, T \tag{3}$$

$$\sum_{i \in F_p} \sum_{r=0}^{C_i} x_{ij(t-r)} \leq 1, \; p = 1, \ldots, NF, \quad j = 1, \ldots, K, \quad t = 1, \ldots, T \tag{4}$$

$$\sum_{i=1}^{N} \sum_{t=1}^{T} C_i \cdot x_{ijt} = T, \quad j = 1, \ldots, K \tag{5}$$

$$x_{ijt} = \{0, 1\}, \quad i = 1, \ldots, N, \quad j = 1, \ldots, K, \quad t = 1, \ldots, T, \tag{6}$$

in which $t - r \leq 0$ then replace with $t - r + T$.

The objective-function (1), to be minimized, is a way to measure the proliferation of pests between the cultures, throughout the planting area, considering the T periods of planning. The closer the cultures of the same botanical family of plants are, greater will be the proliferation, to the detriment of a minor proliferation if the cultures of the same family are far to each other. Objective-function (2) aims to maximize the profit of the planting schedule performed. Note that these goals are conflicting with each other, since, if this function is maximized, only the crops that give more profit will be selected to be planted, that is, the variability between several cultures in planting will be reduced. On the other hand, if z_1 is minimized, crops of different families and of lower profit should be chosen, thus resulting in a lower total profit.

Constraints (3) prevent that there is planting overlapping, that is, for a new crop to be inserted in the same plot, it is necessary to pass C_i periods of its cultivation to have its insertion. Constraints (4) prevent a crop of the same botanical family to be cultivated in a single plot in consecutive periods. Finally, Constraints (5) ensure that the T periods of the schedule in each plot are used. Conditions (6) is the domain of all the decision variables of this problem.

A proposal of linear version of formulation (1)–(6) is presented in the next subsection.

2.1 Linearization of the z_1 Function

As observed, the objective-function z_1 is binary-quadratic. Can not be proved that z_1 is convex, i.e., being $z_1 = x^T Qx$ where Q is the symmetric matrix of order $N + K + T$, there is no guarantee that Q positive defined. These inconvenient cause difficulties in the integer quadratic programming solvers. In the attempt to use integer linear programming to optimize its escalarization, we adopted a procedure of integer programming modeling techniques to linearize it as follows: we replaced the product $x_{ijt} \cdot x_{\bar{i}\bar{j}t}$ by $z_{i\bar{i}j\bar{j}t}$ and we used the linear constraints:

$$x_{ijt} + x_{\bar{i}\bar{j}t} - z_{i\bar{i}j\bar{j}t} \leq 1, \tag{7}$$

$$z_{i\bar{i}j\bar{j}t} \leq x_{ijt}, \tag{8}$$

$$z_{i\bar{i}j\bar{j}t} \leq x_{\bar{i}\bar{j}t}, \tag{9}$$

for all $i = 1, \ldots, N, \bar{i} = 1, \ldots, N, j = 1, \ldots, K-1, \bar{j} = j+1, \ldots, K, t = 1, \ldots, T$ where $z_{i\bar{i}j\bar{j}t} \geq 0$ are new variables.

The linear version for Model (1)–(6) is presented in (10)–(15), where z_1^l is z_1 linearized. Our intention is to compare the performance of ε−constrained considering the nonlinear formulation (1)–(6) and linear one. Note that the model grows in size, both in number of variables and constraints. The details of the scalarization method is present in the next section.

$$\text{minimize} \quad z_1^l = \sum_{i=1}^{N}\sum_{\bar{i}=1}^{N}\sum_{j=1}^{K-1}\sum_{\bar{j}=j+1}^{K}\sum_{t=1}^{T} P_{i\bar{i}} \cdot V_{j\bar{j}} \cdot z_{i\bar{i}j\bar{j}t} \tag{10}$$

$$\text{maximize} \quad z_2 = \sum_{i=1}^{N}\sum_{j=1}^{K}\sum_{t=1}^{T} l_i \cdot area_k \cdot x_{ijt} \tag{11}$$

$$\text{subject to} \quad \text{constraints (3)–(5)} \tag{12}$$

$$\text{constraints (7)–(9)} \tag{13}$$

$$x_{ijt} = \{0, 1\}, \quad z_{i\bar{i}j\bar{j}t} \geq 0, \tag{14}$$

$$i = 1, \ldots, N, \quad \bar{i} = 1, \ldots, N, \quad j = 1, \ldots, K-1, \quad \bar{j} = j+1, \ldots K, \quad t = 1, \ldots, T. \tag{15}$$

3 The ε−Constrained Method

The theoretical results related to this method is presented in [2]. Thus, to solve this bi-objective problem and to obtain Pareto optimal solutions, we considered the function z_1 as the objective to be minimized and set the z_2 as constraint. According the previous authors, the optimal solution of the constrained problem is Pareto optimal for the multi-objective formulation.

In this case, it was necessary to define a range for z_2, avoiding solving idle and infeasible mono-objective problems. As the z_2 function takes only integer values (l_i and $area_j$ can be converted into integers numbers), it was more convenient consider this function as constraint rather than z_1. This justifies the choice of z_1 as the objective of constrained problems (18) or (19).

In this procedure, we used the linear formulation of our problem to guarantee the optimality of these solutions. In order to determine the range of this function, firstly we solved the following problem:

$$\begin{aligned} &\text{minimize} \quad z_1^l \\ &\text{subject to} \quad \text{constraints (12)–(15).} \end{aligned} \tag{16}$$

Let $\mathbf{x}_1^*$ the optimal solution of previous problem. According to [4], if $\mathbf{x}_1^*$ is unique, then we have a Pareto optimal solution for proposed problem (called *left lexicographical solution*). The *left lexicographical point*—z^{left} is defined by the image by z_1 and z_2 of $\mathbf{x}_1^*$ in the criteria space, i.e., $z^{\text{left}} = (z_1(\mathbf{x}_1^*), z_2(\mathbf{x}_1^*)) = (z_1^-, z_2^-)$.

The next step is to determine the other lexicographical solution (called *right lexicographical*), $\mathbf{x}_2^*$. To do so, it is enough to solve the following scalar problem:

$$\begin{aligned} &\text{minimize} \quad z_2 \\ &\text{subject to} \quad \text{constraints (12)–(15).} \end{aligned} \tag{17}$$

With lexicographical solutions, its possible to define a range for z_2 function. The lowest value for z_2 is $z_2^- = z_2(\mathbf{x}_2^*)$ (when the profitability is minimal) whereas the highest value is $z_2^+ = z_2(\mathbf{x}_1^*)$ (when the profitability is maximal). The *right lexicographical point* (z^{right}) is defined as the image of $\mathbf{x}_2^*$ applied in z_1 and z_2 in criteria space, i.e., $z^{\text{right}} = (z_1(\mathbf{x}_2^*), z_2(\mathbf{x}_2^*)) = (z_1^+, z_2^+)$.

The nonlinear formulation of the constrained problem, is given by

$$\begin{aligned} &\text{minimize} && z_1 \\ &\text{subject to} && \text{constraints (3)–(6)} \\ & && z_2 \geq \varepsilon \end{aligned} \tag{18}$$

whereas the linear version can be written as

$$\begin{aligned} &\text{minimize} && z_1^l \\ &\text{subject to} && \text{constraints (12)–(15)} \\ & && z_2 \geq \varepsilon \end{aligned} \tag{19}$$

where, in both cases, $z_2 \in [z_2^-, z_2^+]$. Different solutions can be obtained attributing different values for ε or according the decision marker's preference.

As the solver implemented by CPLEX to solve the mono-objective problems are distinct, we don't expect the same solution for a given value for ε, as well as the computational CPU time to get a given tolerance to the optimal solution. In the next section, we highlight these results and present some Pareto optimal solutions using mosaics.

4 Computational Results

The computational experiments to this problem were implemented in Julia language 0.7.4 version using the package JuMP (see in [2]) in an Intel Core i7 with 8 GB of RAM. Table 1, presents how the 27 semi-random instances generated that we considered validating and compare the linear and quadratic mathematical models. The number of variables and constraints in each formulation are also presented, for all instances. The ones were generated combining different values of N, K and T. In the same table, we present the number of variables and constraints that the formulations determined in the quadratic (1)–(6) and linear model (10)–(15).

We present in Table 2 the results for objective vectors considering the left and right lexicographical points of the Pareto frontier (denoted by $(z_1^-, z_2^-)^T$ and $(z_1^+, z_2^+)^T$ respectively). The function z_2 is measured in U\$. These points were determined by using the linear formulation until we obtain optimal solutions for Problems (16) and (17). The CPU time was omitted. These results allow us to see the amplitude of the objectives functions and emphasize the aspect of conflicting objectives of this problem. For example, in class 23, consider the solutions with minimum proliferation

Table 1 Characteristics of instances, number of variables and constraints of quadratic and linear formulations

Instance	Parameters			Linear formulation		Quadratic formulation	
	N	K	T	# Variables	# Constraints	# Variables	# Constraints
1	5	6	6	2430	2400	180	150
2	5	6	12	4860	4794	360	294
3	5	6	24	9720	9582	720	582
4	5	9	6	5670	3825	270	225
5	5	9	12	11340	7641	540	441
6	5	9	24	22680	15273	1080	873
7	5	12	6	10260	5250	360	300
8	5	12	12	20520	10488	720	588
9	5	12	24	41040	20964	1440	1164
10	10	6	6	9360	9222	360	222
11	10	6	12	18720	18438	720	438
12	10	6	24	37440	36870	1440	870
13	10	9	6	22140	14733	540	333
14	10	9	12	44280	29457	1080	657
15	10	9	24	88560	58905	2160	1305
16	10	12	6	40320	20244	720	444
17	10	12	12	80640	40476	1440	876
18	10	12	24	167280	80940	2880	1740
19	20	6	6	36720	36402	720	402
20	20	6	12	73440	72768	1440	798
21	20	6	24	146880	145590	2880	1590
22	20	9	6	87480	58203	1080	603
23	20	9	12	174960	116397	2160	1197
24	20	9	24	349920	232785	4320	2385
25	20	12	6	159840	80004	1440	804
26	20	12	12	319680	159996	2880	1596
27	20	12	24	639360	319980	5760	3180

pests (presented in Fig. 2) and maximum profit (presented in Fig. 4). While the profit in the second one is twice higher, the proliferation pests increases 20 times.

Consider the class 5 and 23, whose unique difference is the number of different crops. Note that the minimum value for z_1 in class 5 is 4.807, nearly 3 times higher considering the class 23. This means that the greater the variety of crops available, the model is more flexible to determine a calendar with the greatest number of varieties, thus reducing the proliferation of diseases and pests. On the other hand, the number of different crops slight modify the profit of schedules (in this example, it was 4%).

Comparing the classes 9 and 18, both have $K = 12$ lots and $T = 24$ periods of cultivate, but with $N = 5$ and $N = 10$ cultures, respectively. In these classes, we

Table 2 Computational results of three objective vectors from 27 classes

Classe	Lexicographical values				Linear			Quadratic		
	z_1^-	z_1^+	$z_2^- \cdot 10^{-3}$	$z_2^+ \cdot 10^{-3}$	$\bar{z}_1$	$\bar{z}_2 \cdot 10^{-3}$	CPU (sec.)	$\bar{z}_1$	$\bar{z}_2 \cdot 10^{-3}$	CPU (sec.)
1	0.113	9.962	1.193	1.710	1.518	1.480	106.0	1.518	1.518	12.0
2	0.256	19.920	2.027	3.420	8.002	3.250	104.0	8.004	2.783	35.0
3	3.486	39.840	4.519	6.840	8.540	6.348	112.0	8.960	5.250	107.0
4	0.297	17.220	1.789	2.560	9.882	2.410	465.0	10.910	2.356	28.8
5	4.807	34.440	3.179	7.554	7.235	5.038	107.0	10.340	3.987	94.4
6	10.820	68.880	6.510	10.260	*	*	*	17.120	7.683	434.0
7	0.504	24.590	2.385	3.420	17.451	3.420	156.0	18.296	3.195	30.7
8	6.735	49.130	4.148	6.840	*	*	*	24.119	5.503	250.4
9	20.750	98.310	8.894	13.680	*	*	*	23.350	10.760	1630.0
10	0.040	9.960	0.770	1.760	2.924	2.083	97.0	4.560	2.370	30.0
11	0.550	19.920	2.020	3.520	*	*	*	4.173	3.772	412.0
12	0.904	39.840	4.300	7.040	*	*	*	8.565	7.502	862.0
13	0.134	17.220	1.162	2.640	*	*	*	3.635	2.769	313.0
14	0.875	34.440	3.030	5.280	*	*	*	7.083	5.610	1037.0
15	5.554	68.900	6.410	10.560	*	*	*	28.132	13.643	987.0
16	0.175	24.593	1.562	3.520	*	*	*	6.462	3.958	753.0
17	2.810	49.150	4.060	7.040	*	*	*	15.290	8.236	1100.0
18	17.020	98.340	8.620	14.080	*	*	*	46.190	18.790	1880.0
19	0.029	9.150	0.730	1.970	*	*	*	2.243	1.994	350.0
20	0.449	19.920	1.668	3.560	*	*	*	3.696	3.680	1028.0
21	2.290	39.840	3.880	7.120	*	*	*	*	*	*
22	0.133	14.020	1.160	2.900	*	*	*	2.340	2.616	982.0
23	1.810	36.11	3.654	7.631	*	*	*	9.125	5.894	3513.0
24	5.700	68.880	5.843	10.680	*	*	*	*	*	*
25	0.206	20.550	1.550	3.870	*	*	*	3.956	3.560	1437.0
26	1.710	49.140	3.440	7.120	*	*	*	12.331	7.729	3658.0
27	9.060	98.320	7.910	14.240	*	*	*	*	*	*

have the biggest infestation of pests along the planning horizon, due mainly, the maximum values for K, T and $N \leq 10$. On the other hand, the class 19 had the lowest proliferation due to $N = 20$ and $K = T = 6$.

We employed the $\varepsilon-$constrained to find a "*mixed*" solution for this problem, i.e., we determined a Pareto optimal solution in each class using $\varepsilon = \bar{\varepsilon} = \frac{1}{2}(z_2^- + z_2^+)$. This solution trends has an objective vector as average of lexicographical vectors. The symbol (*) indicated that the CPLEX did not determine any feasible integer solution in 3600 s. As we can see, in linearized formulation, CPLEX determined a sub-optimal solution for the constrained problem to only 7 instances, while in the quadratic one, CPLEX determined sub-optimal solution in 24 instances. The *mixed* solutions determined by the linear and quadratic formulations are, in general, different due to the algorithms involved, criteria stopping, etc. The three instances

<table>
<tr><th>Plot\Per</th><th>1</th><th>2</th><th>3</th><th>4</th><th>5</th><th>6</th><th>7</th><th>8</th><th>9</th><th>10</th><th>11</th><th>12</th></tr>
<tr><td>1</td><td>19</td><td colspan="3">2</td><td colspan="3">13</td><td colspan="3">9</td><td colspan="2">19</td></tr>
<tr><td>2</td><td colspan="2">17</td><td colspan="4">14</td><td colspan="3">19</td><td colspan="3">9</td></tr>
<tr><td>3</td><td>8</td><td colspan="3">2</td><td colspan="4">14</td><td colspan="3">9</td><td>8</td></tr>
<tr><td>4</td><td colspan="3">13</td><td colspan="2">6</td><td colspan="4">14</td><td colspan="3">2</td></tr>
<tr><td>5</td><td colspan="2">1</td><td colspan="3">7</td><td colspan="3">19</td><td colspan="3">9</td><td>20</td></tr>
<tr><td>6</td><td>7</td><td colspan="2">15</td><td colspan="3">19</td><td colspan="4">14</td><td colspan="2">7</td></tr>
<tr><td>7</td><td>13</td><td colspan="3">7</td><td colspan="3">9</td><td colspan="3">19</td><td colspan="2">13</td></tr>
<tr><td>8</td><td>2</td><td colspan="4">14</td><td colspan="3">7</td><td colspan="2">17</td><td colspan="2">2</td></tr>
<tr><td>9</td><td colspan="3">14</td><td colspan="3">9</td><td colspan="3">7</td><td colspan="2">17</td><td>14</td></tr>
</table>

Fig. 2 The representation of the solution of class 23 minimizing z_1 (minimum proliferation)

<table>
<tr><th>Plot\Per</th><th>1</th><th>2</th><th>3</th><th>4</th><th>5</th><th>6</th><th>7</th><th>8</th><th>9</th><th>10</th><th>11</th><th>12</th></tr>
<tr><td>1</td><td colspan="3">7</td><td>3</td><td colspan="2">11</td><td colspan="2">8</td><td>20</td><td colspan="2">4</td><td>10</td></tr>
<tr><td>2</td><td colspan="2">15</td><td colspan="2">8</td><td colspan="2">5</td><td>10</td><td>3</td><td colspan="2">1</td><td colspan="2">18</td></tr>
<tr><td>3</td><td>3</td><td colspan="2">1</td><td>3</td><td colspan="2">1</td><td>3</td><td colspan="2">1</td><td>3</td><td colspan="2">1</td></tr>
<tr><td>4</td><td>1</td><td>3</td><td colspan="2">1</td><td>3</td><td colspan="2">1</td><td>3</td><td colspan="2">1</td><td>3</td><td>1</td></tr>
<tr><td>5</td><td colspan="2">13</td><td colspan="3">2</td><td>16</td><td>20</td><td colspan="2">18</td><td colspan="2">8</td><td>13</td></tr>
<tr><td>6</td><td colspan="2">1</td><td colspan="2">12</td><td>16</td><td colspan="3">9</td><td colspan="2">4</td><td colspan="2">8</td></tr>
<tr><td>7</td><td>17</td><td colspan="2">1</td><td colspan="2">11</td><td>16</td><td>3</td><td colspan="2">12</td><td>20</td><td>10</td><td>17</td></tr>
<tr><td>8</td><td>1</td><td colspan="3">13</td><td>20</td><td colspan="3">9</td><td colspan="2">17</td><td>3</td><td>1</td></tr>
<tr><td>9</td><td>3</td><td colspan="2">1</td><td>3</td><td colspan="2">1</td><td>3</td><td colspan="2">1</td><td>3</td><td colspan="2">1</td></tr>
</table>

Fig. 3 The representation of the solution of class 23 with $\varepsilon = \frac{1}{2}(z_2^- + z_2^+)$ (mixed solution)

Plot\Per	1	2	3	4	5	6	7	8	9	10	11	12
1	3	10	3	10	3	10	3	10	3	10	3	10
2	3	10	3	10	3	10	3	10	3	10	3	10
3	3	10	3	10	3	10	3	10	3	10	3	10
4	10	3	10	3	10	3	10	3	10	3	10	3
5	3	10	3	10	3	10	3	10	3	10	3	10
6	3	10	3	10	3	10	3	10	3	10	3	10
7	3	10	3	10	3	10	3	10	3	10	3	10
8	3	10	3	10	3	10	3	10	3	10	3	10
9	3	10	3	10	3	10	3	10	3	10	3	10

Fig. 4 The representation of the solution of class 23 with maximum profitability

in which the quadratic failed to find a feasible integer solution in 3600 s were those with a larger number of variables and constraints. With the non-convex quadratic function (1), CPLEX ends its search in a point that is a local optimum.

Consider the three solutions in class 23, represented by the Figs. 2, 3 and 4, whose objective z_1 is 1.810, 9.125 and 36.11, i.e., an increase of 400% and 300%, respectively. Analyzing the z_2 function, we have $3.654 \cdot 10^3$, $5.894 \cdot 10^3$ and $7.631 \cdot 10^3$, an increase of 61% and 29%, respectively. These percentages, obviously, does not consider the dimension of z_1 and z_2. Then an improvement in z_2 promotes a large changes in objective z_1.

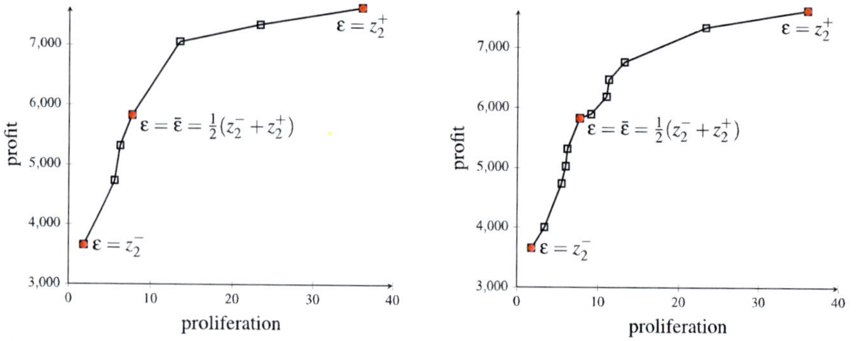

(a) Non-dominated points to instance 23 using 5 attributions to ε

(b) Non-dominated points to instance 23 using 10 attributions to ε

Fig. 5 Other non-dominated points determined for instance 23

We note that the solution with minimum proliferation uses the 12 crops in the calendar, while in the mixed solution uses 17 crops and, finally, in solution with maximum profit we have just two cultures (rotation of crops 3 and 10 because give the highest profit per hectare). In last case, the proliferation is high because neighboring lots (with common edge) have the same crops in each period. Thus, z_1 decrease as the multi-objective model eliminates the same cultures in adjacent lots in each period.

Although the mixed solution uses 5 cultures more, we observe that the average of cycle duration of the inserted cultures is lower compared to solution with minimum proliferation. The model decides to plant cultures with life-cycles shorter in a intermediate solution. In this case, we have 24 times the cultures of life-cycle unitary be planted while, in other solution, we just have cultures of life-cycle non-unitary.

We present in Fig. 5, two particular simulations for class 23, showing the ability of $\varepsilon-$constrained in determine other Pareto optimal solutions and different scenarios of cultivate. Figure 5a, b illustrate 5 and 10 assignments for ε values, uniformly distributes in the range $[z_2^-, z_2^+] = [3654.00, 7631.50]$. These solutions were obtained by the quadratic formulation (we limited the CPLEX to 7200 s for solve each mono-objective problem). We emphasize the non-dominated points determined by the lexicographical solutions and with $\varepsilon = \bar{\varepsilon}$.

5 Conclusions

The article showed the application of an exact technique of mathematical programming to determine potentially efficient solutions crop rotation problem, implemented by the CPLEX solver. Both the quadratic and linear formulations are tested. The study illustrated, through computational tests in instances of real dimensions, that the exact method has is more efficient, from computational viewpoint, when we use the quadratic formulation. However, the solutions are sub-optimal, since the first objective-function is not convex.

In short, the mathematical model proposed, which aims to establish a way of sustainable vegetable production compromised with the profit of the agriculture producer, as well as the solution method proposed, are applicable in real cases, helping decision-makers in choosing different alternatives of agricultural production in their farms and contributing to the advance of knowledge in the field of Operational Research in a sustainable environment.

Future researches are also proposed with this study. To solve larger instances and those with no solution in reasonable CPU time, we intend to propose a genetic algorithm to provide Pareto sub-optimal solutions with low computational cost. Thus, a comparative performance between the exact and heuristic method can be made.

Acknowledgements The authors thank the Brazilian institution UTFPR, EDITAL DIRPPG-AP 01/2019, PROEPE (UNESP) and FUNDUNESP. In addition, we thank the Federal Technological University of Parana for the support of this research and by the translation services provided.

References

1. Aliano Filho, A., de Oliveira, Florentino H., Vaz Pato, M.: Metaheuristics for a crop rotation problem. Int. J. Metaheuristics **3**(3), 199–222 (2014)
2. Aliano Filho, A., de Oliveira Florentino, H., Pato, M.V. et al. Exact and heuristic methods to solve a bi-objective problem of sustainable cultivation. Ann Oper Res (2019). https://doi.org/10.1007/s10479-019-03468-9
3. Havlin, J., Kissel, D., Maddux, L., Claassen, M., Long, J.: Crop rotation and tillage effects on soil organic carbon and nitrogen. Soil Sci. Soc. Am. J. **54**(2), 448–452 (1990)
4. Miettinen, K.: Nonlinear Multiobjective Optimization, International Series in Operations Research and Management Science, vol 12. Kluwer Academic Publishers, Dordrecht (1999)
5. Santos, L.M., Munari, P., Costa, A.M., Santos, R.H.: A Branch-price-and-cut method for the vegetable crop rotation scheduling problem with minimal plot sizes. Eur. J. Oper. Res. **245**(2), 581–590 (2015). https://doi.org/10.1016/j.ejor.2015.03.035, http://www.sciencedirect.com/science/article/pii/S0377221715002428
6. Santos, L.M.R., Costa, A.M., Arenales, M.N., Santos, R.H.S.: Sustainable vegetable crop supply problem. Eur. J. Oper. Res. **204**(3), 639–647 (2010)
7. Santos, L.M.R., Michelon, P., Arenales, M.N., Santos, R.H.S.: Crop rotation scheduling with adjacency constraints. Ann. Oper. Res. **190**(1), 165–180 (2011)
8. Snapp, S., Swinton, S., Labarta, R., Mutch, D., Black, J., Leep, R., Nyiraneza, J., O'neil, K.: Evaluating cover crops for benefits, costs and performance within cropping system niches. Agron. J. **97**(1), 322–332 (2005)
9. West, T.O., Post, W.M.: Soil organic carbon sequestration rates by tillage and crop rotation. Soil Sci. Soc. Am. J. **66**(6), 1930–1946 (2002)

In-House Logistics Operations Enhancement in the Automobile Industry Using Simulation

Rodrigo Macedo, Fábio Coelho, Susana Relvas, and Ana Paula Barbosa-Póvoa

Abstract In a fast-paced and synchronised context for manufacturing industries, an efficient material handling of parts is crucial. Therefore, this industry requires reliable and efficient internal logistics operations that enable efficient production strategies, such as an assembly line powered by a decentralized storage support area—logistics supermarket. Agile and flexible decision-making raises the need to plan all plant operations as well as to control them. To support such planning this work proposes a simulation-based decision support tool for the operation of a manufacturing supermarket. It analyses the activity of order picking and line feeding in the logistics supermarket where factors such as the speed of the pickers or the number of AGV's are explored. This is done using the Simio—simulation modelling framework based on intelligent objects. The model was validated using a real case study.

Keywords In-house logistics · Simulation · Assembly line feeding · Material handling · Manufacturing supermarket

1 Introduction

The increasing competitiveness imposes manufacturers the need for efficient operational strategies. In order to stay competitive, original equipment manufacturers (OEM's) are now offering a value-added service, where customers can choose specific parts in the vehicle [1]. Consequently, each vehicle at the assembly line is tailored, with specific parts to be installed and, therefore, mixed-model assembly line (MMAL's) have been adopted in the industry [2]. To guarantee flexibility in-house activities are a crucial but the complexity of the associated process concerning the

R. Macedo · F. Coelho · S. Relvas (✉) · A. P. Barbosa-Póvoa
CEG-IST, Instituto Superior Técnico, University of Lisbon, Av. Rovisco Pais, 1049-001 Lisbon, Portugal
e-mail: susana.relvas@tecnico.ulisboa.pt

F. Coelho
e-mail: fabiocoelho@tecnico.ulisboa.pt

S. Relvas et al. (eds.), *Operational Research*, Springer Proceedings in Mathematics & Statistics 374, https://doi.org/10.1007/978-3-030-85476-8_4

material handling at the shop floor has increased exponentially. Different assembly line feeding policies, following just-in-time (JIT) philosophy, are being tested with the intention of securing a steady flow in production, decreasing operational costs, ensuring high quality standards [3] and allowing for more flexible, faster and reliable logistics processes [4]. In this set of increased complexity decision supporting tools to help in-house logistics related operations decisions are required.

We develop a simulation-based decision-making tool to analyse order picking operations in a manufacturing supermarket, as well as the related feeding activities. Specifically, in this work, we perform sensitivity analyses on several parameters of the model, which allows recognizing how the system responds to different situations that allow to further improve the operation. This is done based on a real case study. Also, we show the importance of simulation usage in complex manufacturing contexts.

Section 2 presents a literature review based on manufacturing supermarket. In Sects. 3 and 4, the problem description and its implementation on Simio are explained. Then, the model validation is described and results are presented in Sect. 5. Finally, conclusions are provided in Sect. 6.

2 Literature Review

According to [5], logistics systems are always open and exchanging information and material with its surroundings, but the boundaries can be different depending on the perspective. In-house logistics is one of these perspectives, where the boundaries in which the exchange happens, correspond to the physical limits of the company [6], meaning that the logistic process occurs within the company walls.

The main fields of in-house logistics are warehousing, transportation and production line organization [7], which are present in manufacturing industries based in assembly line processes. Regarding warehousing, there are two main options to choose from, which are the use of centralized or decentralized storage points. Traditionally, central storage points were always the option, but manufacturing facilities today are so large that this is no longer a viable possibility. The second function is transportation, whose task is the transportation of material inside the plant, after being delivered by the suppliers. Finally, the third function is production line organization or more specifically line side presentation, that tackles the problem of how bins (or other objects able to carry parts) are placed in the border-of-line (BoL). This is one of the most relevant functions in in-house logistics since it is in the assembly line where most value is created. Thus, it is necessary a continuous improvement to enable a steady flow in production mostly through ergonomic factors, which enable better working conditions for assembly line workers [8]. Today, the complexity relative to the production of vehicles becomes so much more complex than concepts like part logistics and assembly line feeding systems following JIT philosophy (which aims at synchronizing the supply of parts with their demand [9]) started to be adopted by OEM's around the globe [2] allowing more flexible, reliable and faster logistics

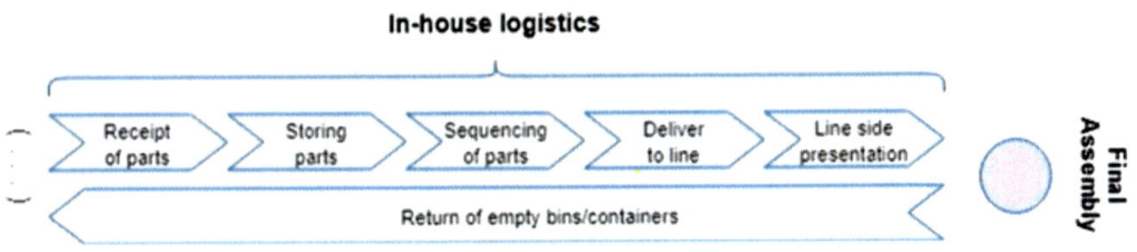

Fig. 1 In-house part logistics (adapted from [18])

processes [4]. In the automotive industry, the standard procedure in which parts are treated is depicted as in Fig. 1.

The main goal is to have a secure and steady feeding of parts to the assembly line, avoiding stock-outs and, consequently, line-stoppage and idleness of workstations and workers [3]. The order in which these processes are executed is not rigid and can vary depending on the feeding policy applied. Still, it is important to notice that all of these processes are interdependent [10], as decisions upstream of the logistics part process will influence decisions downstream and vice versa. Parts feeding systems have become an intrinsic part of today's assembly line operations. These systems assure that parts are available where and when they are needed in diverse assembly operations. The delivery of components or sub-assemblies to the assembly line is done in pre-determined quantities and sequences, in specific containers which are called kits, following a JIT supply of parts to the assembly line. A kit, as defined by [11], is a "specific collection of components and/or sub-assemblies that together (i.e., in the same container) support one or more assembly operations for a given product or shop order", where a component is something that cannot be divided and sub-assemblies is a construction that is composed of two or more components. In order to assemble the kit, it is required a prior operation to line feeding which concerns the sequencing of parts, much similar to order picking. After completed, kits can be stored in a special type of rack that can be fixed or movable and store various kits. The content of kits are constrained by maximum weight, volume and the order in which kits are assembled and delivered to the line, according to a pre-determined assembly line production schedule [12, 13]. Once the kit is fulfilled it needs to be transported to the correct point-of-fit (POF)[1] in the assembly line. The delivery is normally done by tow-trains (automated or manual) that perform milk-runs (i.e., loops) inside the plant. When delivered at the assembly line there are two types of kits: (i) those who follow the assembly line products (i.e., travelling kits), so that the kit parts are used in more than one workstation [8], and (ii) those that remain at a workstation until they are depleted (i.e., stationary kits) [11].

Some works have been done regarding the subject such as those by [11, 13–16]. Most of these works are done via comparison with other feeding policies such as line stocking. The conclusions of these various works are common among them. In summary kitting leads to lower stock levels at the line as well as diminish work-in-progress (WIP) and more simple material flow at the shop floor. Kitting is best used

[1] Point-of-fit (POF)—Location where the part is to be placed or consumed.

when dealing with mass customization, and low standardised parts with small to medium physical dimensions. Its main disadvantages are the extra material handling needed (and associated costs) and the chance of having errors in the building of kits which can lead to line stoppages. Another important concept related to the kitting is where the kitting is fulfilled. JIT-Supermarkets are decentralized intermediate storage points for components and sub-assemblies, that are strategically located at nearby line segments which they serve [7], allowing for flexible and frequent small-lot deliveries. Normally, replenished through tow-trains or more capacitated transportation modes, from a central storage point. They are then delivered to the line segments also via tow-trains or automated guided vehicles (AGV). This concept has also been a motif for attention as they have become widely used in the industry. Some works, that were reviewed regarding the subject (e.g., [2, 9, 14] mostly focus on four main topics which are: (1) localization of the supermarket; (2) routing of the tow trains/AGV; (3) scheduling and (4) loading. For clarification topic (3) refers to the delivery schedule of tow trains/AGV to the assembly line and (4) to the decision about lot sizes and types of parts for tow trains/AGV transport each time.

Lastly, given the similarity between kitting and order picking some works were reviewed, having as focus pickers-to-parts systems using batching which is the collection of multiple orders at the same time. One of the most relevant work for our study was accomplished by [16]. They state that the most important aspects regarding kit elaboration efficiency are higher picking density, size of area, racks for storage of packages (number of levels), area layout, moving or stationary kit rack, storage package (design of a part), batch size, kit rack (how many racks, design) and finally information system (e.g., picking list, pick-by-light). Some of its findings show that a higher pinking density negatively correlates to the time spent picking parts, meaning the higher the density the shorter is the time needed to collect apart. The information system also plays a significant role as some require more administrative time than others (e.g., in order to activate the pick-by-light system the picker must first scan a paper containing the requested kits). Also, the packaging of the stored parts if not readily accessible could decrease efficiency (e.g., parts wrapped in plastic film, among others).

3 Problem Description

The real case study supporting this work is a vehicle manufacturer based in Portugal, which is currently introducing a new vehicle model and suffering capacity limitations. The challenge focuses on the engine assembly line, where the transportation of parts goes from a decentralized storage point (also known as supermarket) to the assembly line by a combination of workers and automated guided vehicles (AGV's). Here it is observed a low efficiency regarding workers as the arrival of AGV's at the collection point is not synchronised with the completion of parts collection tours. Thus, with the aim of improving the operation, a simulation model is developed where different operational strategies are experimented. The problem focused on a section of the

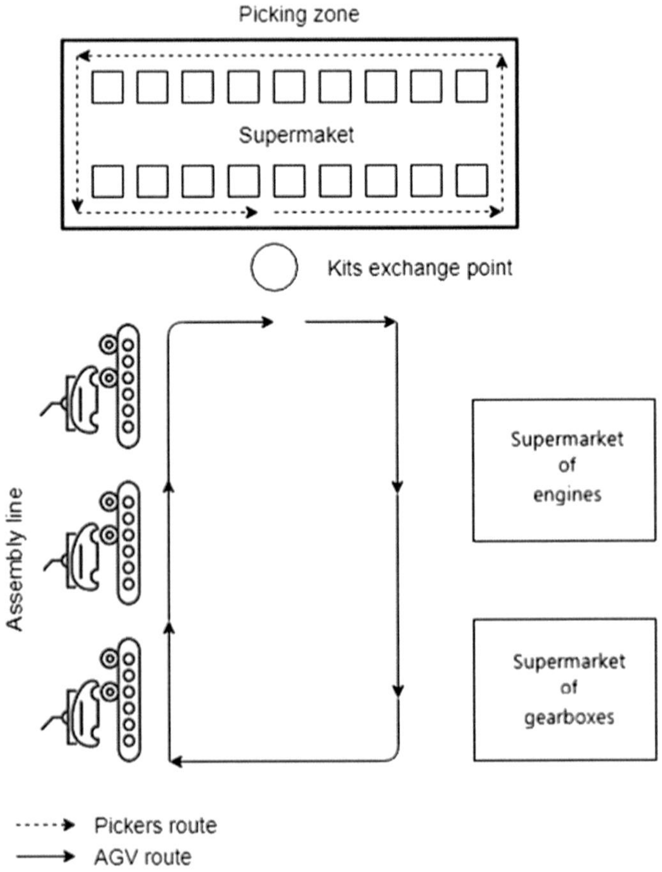

Fig. 2 Overview of the system under study

assembly line dedicated to engines. It is composed of the main supermarket where kits are prepared by four human pickers, eleven AGV's that act as tow trains to collect and deliver kits/parts, plus two more supermarkets that store engines and gearboxes and, finally, diverse workstations where the kits and parts are to be consumed. A general view of the process can be observed in Fig. 2.

The process starts with the pickers collecting parts at the supermarket in order to fulfil the kits, which when completed, are unloaded manually into an AGV at the exchange point. From here the AGV passes through the other two supermarkets collecting an engine and a gearbox. Afterwards, it waits at the entrance of the assembly line for the correct time window to deliver the first set of parts in order to respect the factory sequencing order (since all vehicles are tailored). The kits are unloaded to another type of AGV that act as the assembly line and which will from now on be referred to as production-AGV (PAGV). After it delivers the first set of parts, the AGV proceeds to the subsequent workstation until all kits and parts are depleted.

At each workstation, the AGV has to wait for an assembly line worker to finish its current work so he can retrieve the items manually.

This type of systems has many stochastic characteristics. Uncertainties referring to the type of engine to be assembled (i.e., more or less complex), sporadic events such as a missing part or an employee being slower than usual, for example, directly influence time and consequently increase the difficulty of assertiveness when interpreting results. Given these reasons, a discrete-event simulation model integrating all these problems is developed. Simulation is adequate to address complex systems and test different scenarios or configurations while analyzing the outcomes through a systematic evaluation of operational indicators.

4 Model Development

The activity cycle diagram, shown in Fig. 3, was developed. This describes the whole system where two types of entities were considered: (i) temporary—entities that enter and leave the system (e.g., kit types) and (ii) permanent—entities that always remain in the system (e.g., humans and AGV's).

As a first step for developing the model, the relevant data was gathered. The main sources for this collection were in-situ observations and the Enterprise Resource System (ERP). The data mainly consisted of objects independent behaviours, picker and AGV routes and times, manpower costs, line stoppage costs, picking locations, various kits with respective sequences and real-life probabilities of demand, objects physical dimensions, distances, physical layouts, work schedules and speeds. These data were worked in order to be represented by statistical distributions that serve as input to the simulation model.

With all the foundations firmly defined, as well as all the relevant data gathered, the computational step was accomplished. For that, Simio simulation software was chosen as it is simple to present the model to decision-makers once the model looks

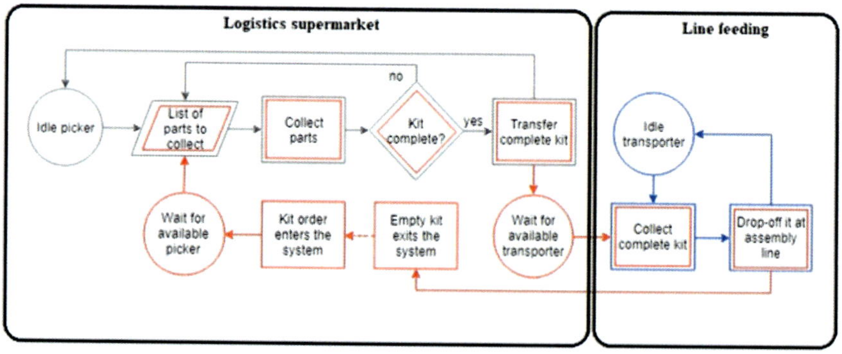

Fig. 3 Activities cycle diagram of the model

Table 1 Modelling entities

Real system entities	Modelled as	Description
Empty JIS-bin	Entities sequentially generated by a "source" object	Entities that are processed in the supermarket
Human picker	"worker"	Its function is to fulfil the orders according to each kit type
Product locations	Several "combiners"	Represents the storage locations of the different products in the supermarket
Products	Entities generated by a "source" object	Entities that are waiting for picking in the storage locations
Kit	"model entity"	Represents a fulfil order ready to feed the assembly line
AGV	"vehicle"	Its function is to perform the assembly line feeding
Assembly line	Several "workstations"	Represents the working in series

identical to the real world. Through this software, it was possible to input all the data gathered into the model with a high degree of detail. It is important to highlight some assumptions made. The first assumption states that there is no possibility of stock out at the supermarket. Second, no shortcuts in the picking of parts are allowed from the pickers. Third, PAGV is merely considered as a resource for AGV's to be able to unload kits.

As all the data was inputted into the model, the first thing to be modelled was the run time of the simulation which was defined as a full 24-hours day minus lunch and intermediate breaks. Also, objects from the standard library available in Simio were used as well as some custom logic objects to address more specific processes were developed. These "objects" are pre-coded resources that are able to undertake a special task (for readers unfamiliar with Simio, please see [17]). Table 1 presents the considered modelling entities.

Thus, picking locations were modelled through combiners that use the batch function to build the kits in a continuous flow, AGV's and Pickers were programmed using the vehicle and worker objects, respectively. Workstation and the unloading and loading of kits onto the AGV server objects were used. Lastly, for kits entering and leaving the system source and sink objects were used, respectively. Sources were modelled to generate different types of kits based on their real probability. Each specific kit was also modelled to have its own picking locations sequence. Next, each picking location was assigned with processing time. This processing time is based on a probabilistic function gathered which depicts the average collection time for a single part. If it is requested to pick more than one part at a given location, then the processing time would be multiplied by the total quantity to be collected. An average speed of 4 km/h was attributed to the pickers as they have to push a movable kit while stopping at the various picking locations. Custom logic was also inserted

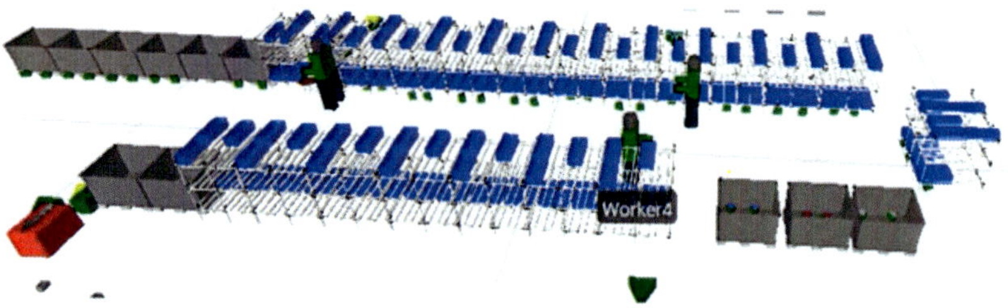

Fig. 4 3D Excerpt of the simulation model

into the model so that pickers did not pass in front of each other so as to not disturb the sequence in which kits need to arrive at the assembly line, applying a first-in-first-out (FIFO) methodology when these entered in any intersection or any path were more than one picker was present. For AGV's a speed of 1.2 m/s was assigned, this velocity is the maximum allowed speed at the factory due to safety reasons. Custom logic was also added to the AGV's which made them always respect the correct sequence when intersections appeared. With the computational modulation complete, it was possible to test several performance indexes related to the system. To have a reliable sample set the simulation underwent 20 independent runs. This allowed not only to have a broader spectrum of different and independent, sets of results but also to have well defined true mean, upper and lower bounds, when applying a 95% confidence interval. An excerpt of the simulation model is presented in Fig. 4.

5 Model Validation and Results Analysis

For model validation, the results obtained were directly compared to the real observations of the factory. For that four main indicators were chosen: (i) tour collection time; (ii) AGV cycle time; (iii) kit orders generation; (iv) engine production per day.

Regarding indicator (i), an average time of 208 s per tour collection was observed versus 228 s with standard deviations of 1 s and from 30 s for real life observation. For indicator (ii), the simulated AGV cycle time was 16 min versus 15.5 min with standard deviations of 17 s versus 57 s, respectively. The kit generation was also deemed reliable as the generated quantities throughout a whole day presented values in accordance with their associated real probability. Lastly, average engine production per day was 894 from the model versus 905 from reality with a standard deviation of 7 regarding the simulation value and 20 regarding reality. The model was deemed reliable and ready for experimentation, since the simulation analyst and experts on the field had reviewed the model results for correctness. In order to be able to analyse the system, some key performance indicators (KPI's) are defined so as to quantify different parameters. These indicators are the following:

Table 2 Current state analysis (95% CI)

	Lower	Average	Upper	Standard deviation
KPI 1 (hours/day)	8.2	8.4	8.5	0.22
KPI 2 (units)	0	0.006	0.0113	0.015
KPI 3 (€/day)	294	302	307	8
KPI 4 (€/day)	6	58	110	113
KPI 5 (€/day)	300	359	416	121
KPI 6 (units/day)	888	894	900	6
KPI 7 (seconds)	207	208	209	1

- KPI1—Picker Idle time—The time that pickers spend waiting for an available AGV at the entrance of the supermarket to exchange kits;
- KPI2—Engines waiting assembly—Total time that engines stay at the beginning of the assembly line waiting to enter the 1st assembly station as there are no kits yet available brought by the AGV, which accounts for a line stoppage;
- KPI3—Picker idle costs—Costs associated with having idle pickers;
- KPI4—Line stoppage costs—Costs associated with having engines accumulated at the entrance of the assembly line. This leads to high costs, as it can lead to full-scale line stoppage. Here a symbolic value is added as it corresponds only to a segment of the assembly line;
- KPI5—Overall costs—As the name implies the overall costs of the system (i.e., KPI3 plus KPI4);
- KPI6—Produced engines—Total production of engines in a single day;
- KPI7—Tour collection time—Total time that a picker takes to collect the requested kits at the supermarket.

Table 2 presents the values for the different key performance indicators. These values are the means resulting from 20 simulation runs. For the width of the true mean, a confidence interval (CI) of 95% was applied. The average per day, considering the whole four pickers, is around 8.4 h of idleness for single picker throughout 24 h with a standard deviation of roughly 0.22 h. According to the statistics provided by the model, there is a 95% chance that the average idle time per run will hardly be less than 8.2 h. This means that for a single picker he spends around 2.5 h idle for each of the three turns leading to an efficiency of only 63.5% per picker. As for the performance indicators regarding AGV's (i.e., KPI2 and KPI4) the results also go in accordance with what is seen in real life.

According to Table 2, the company is losing hypothetically 300 €/day when pickers are concerned. We cannot state that these losses are directly correlated with the final product as the demand is always fulfilled. What we do have is an imbalance regarding manpower in supermarket C3 since the current workforce is higher than it should be. These losses are then a representation of wasted manpower that should be reallocated to another area.

Table 3 Picker speed sensitivity analysis

	Base (4.0 km/h)	Base + 0.5 km/h	Base + 1.0 km/h
KPI 1 (hours/day)	8.38	8.99	9.29
KPI 2 (units)	0.006	0.003	0.002
KPI 3 (€/day)	302	324	334
KPI 4 (€/day)	58	23	18
KPI 5 (€/day)	359	347	352
KPI 6 (units/day)	894	893	901
KPI 7 (seconds)	208	201	195

On the other hand, the company seems to lose around 57 €/day when it comes to line stoppages. How can this be possible if demand is always satisfied? The answer can be found in the model reports, where for the 20 runs the maximum waiting time of a PAGV was 1 min, and the mean average regarding the same 20 runs is 0.5 s with an upper bound of 1 s (95% CI). Knowing this, 57 €/day seems just as unavoidable operational costs and represents a good solution.

For the sensitivity analysis, three factors are chosen to see how variations of these would impact the system. The three factors are picker speed, production rate, and the number of tow trains. For the first two factors, they were chosen due to the fact that they represent high variability/unpredictability in real-life and also, the company wants to study the impact of these factors since some of them are more hypothetical and prevent you from having to test it in the shop-floor. Speed in the model was made constant with a base value of 4 km/h where in reality the velocity of pickers changes given the physical characteristics of the individual, the psychological state and so on. Regarding demand, it is common to see peaks which lead to a higher pressure in the assembly line which can sometimes lead to less smooth operations. Finally, for the tow trains, it seemed that there was a surplus and it would be interesting to see if the factory would be able to cope with production by reducing the number of AGV's. According to Table 3, as the speed of picker increases, we can see that pickers costs are positively correlated as well as the total idle time as expected. However, we can see that with an increment to 5 km/h the total production of engines increases slightly. But in overall increasing picker speed does not lead to any profound impacts in the system and while the line stoppage costs decrease, they are almost overshadowed by the costs resulting from the accumulated costs increased from the picker idleness.

In contrast with picker velocity, increments in demand lead to much greater impacts in the system. From Table 4 all increments in production lead to a greater number when it comes to completed engines in a linear manner. However, the same behaviour is not seen when it comes to line stoppage costs as it increases at a much greater scale leading 5095 €/day in overall costs in the 15% increment scenario. It seems that even if the total production number is increased pickers still have a lot of idleness which leads to the conclusion that the problem does not refer to how

Table 4 Production rate sensitivity analysis

	Base	Base + 5%	Base + 10%	Base + 15%
KPI 1 (hours/day)	8.38	7.79	7.00	6.11
KPI 2 (units)	0.006	0.014	0.002	0.491
KPI 3 (€/day)	302	280	252	220
KPI 4 (€/day)	58	136	963	4875
KPI 5 (€/day)	359	417	1216	5096
KPI 6 (units/day)	894	934	982	1035
KPI 7 (seconds)	208	208	208	208

Table 5 AGV's number sensitivity analysis

	Base (11 units)	Base −2 units	Base −4 units	Base −6 units
KPI 1 (hours/day)	8.38	8.50	8.43	9.13
KPI 2 (units)	0.006	0.043	0.242	29.23
KPI 3 (€/day)	302	306	304	329
KPI 4 (€/day)	58	426	2413	289730
KPI 5 (€/day)	359	732	2717	290059
KPI 6 (units/day)	894	895	900	854
KPI 7 (seconds)	208	208	208	208

quickly kits can be elaborated but probably in the assembly line itself as the current production capacity is not able to deal with such a production rate.

Finally, when it comes to the reduction of the AGV's (Table 5), although it seems that production seems to remain in the average numbers, at least with 9 and 7 AGV's. However, it seems that engines wait longer to commence assembly, which leads to higher costs. It is important to note that in all scenarios line stoppage costs increase greatly due to the high value associated, which is 7.5 €/min. So even if production is not greatly affected having engines in queue leads to high costs. Thus, it can be said that 7 and 9 AGV's are able to cope with production, but with only 5 AGVs lead to production decreases meaning that the requisites needed for the system to operate efficiently are not fulfilled with around 289 729 €/day of line stoppage costs.

6 Conclusions

As a way to cope with the magnitude of parts caused by mass customization of vehicles offered to clients, OEM's have continuously tried to improve, innovate in their in-house logistics operations through the use of new technologies and with the creation of new concepts and methodologies.

In this work, having as basis a real case study, it was possible to grasp the impact of this immensity of parts and how it perpetuates in in-house logistics operations. This was done through the development of a simulation model that allowed the company to study a set of scenarios that were defined as a possible way to answer to the increased system complexity. Critical variables were tested and the results showed that when picker velocity is incremented no meaningful impact is observed in the system, other than the one resulting in a lower collection time while diminishing the number of line stoppages. In the case where the production is incremented it was observed that the problem is not in the supermarket as pickers still have considerable idle time and completed engines always increase. The problem is more related to the processing times at the assembly line that results in a queue when it comes to engines waiting for assembly. Finally, regarding the reduction of AGV's it seems that the breaking is around 5 AGV's and by having only this respective quantity the system is not able to face production demand.

In conclusion, through the developed simulation model was possible to better understand how different KPI's correlate to each other, and how this information could be used to achieve a better overall system performance. Hence the simulation model proved itself as an imperative decision-making tool, which could benefit companies at the short term by providing a better understanding of its operational processes but also in the long term when it comes to the building of new hypothesis and assessing their impacts.

Acknowledgements The authors gratefully acknowledge the financial support from Portugal 2020 framework, under the project POCI-01-0145-FEDER-016418 financed by EU/FEDER through the programme COMPETE2020.

References

1. Coelho, F., Relvas, S., Barbosa-Povoa, A.P.: Simulation of an order picking system in a manufacturing supermarket using collaborative robots. In: Nolle, L., Burger, A., Tholen, C., Werner, J., Wellhausen, J. (eds.) ECMS 2018 Proceedings, pp. 83–88 (2018). https://doi.org/10.7148/2018-0083
2. Emde, S., Boysen, N.: Optimally locating in-house logistics areas to facilitate JIT-supply of mixed-model assembly lines. Int. J. Prod. Econ. **135**(1), 393–402 (2012). https://doi.org/10.1016/j.ijpe.2011.07.022
3. Faccio, M., Gamberi, M., Persona, A., Regattieri, A., Sgarbossa, F.: Design and simulation of assembly line feeding systems in the automotive sector using supermarket, kanbans and tow trains: a general framework. J. Manag. Control **24**(2), 187–208 (2013). https://doi.org/10.1007/s00187-013-0175-1
4. Saaidia, M., Durieux, S., Caux, C.: A survey on supermarket concept for just-in.time part supply of mixed model assembly lines. In: 10ème Conférence Francophone de Modélisation, Optimisation et Simulation, MOSIM 2014 (2014)
5. Jonsson, P.: Logistics and Supply Chain Management. McGraw-Hill Higher Education, UK (2008)
6. Gupta, T., Dutta, S.: Analysing materials handling needs in concurrent/simultaneous engineering. Int. J. Oper. Prod. Manag. **14**(9), 68–82 (1994). https://doi.org/10.1108/01443579410066776

7. Battini, D., Boysen, N., Emde, S.: Just-in-Time supermarkets for part supply in the automobile industry. J. Manag. Control **24**(2), 209–217 (2013). https://doi.org/10.1007/s00187-012-0154-y
8. Sali, M., Sahin, E.: Line feeding optimization for just in time assembly lines: an application to the automotive industry. Int. J. Prod. Econ. **174**, 54–67 (2016). https://doi.org/10.1016/j.ijpe.2016.01.009
9. Golz, J., Gujjula, R., Günther, H.O., Rinderer, S., Ziegler, M.: Part feeding at high-variant mixed-model assembly lines. Flex. Serv. Manuf. J. **24**(2), 119–141 (2012). https://doi.org/10.1007/s10696-011-9116-1
10. Kilic, H.S., Durmusoglu, M.B.: Advances in assembly line parts feeding policies: a literature review. Assem. Autom. **35**(1), 57–68 (2015). https://doi.org/10.1108/AA-05-2014-047
11. Bozer, Y.A., McGinnis, L.F.: Kitting versus line stocking: a conceptual framework and a descriptive model. Int. J. Prod. Econ. **28**, 1–19 (1992). https://doi.org/10.1016/0925-5273(92)90109-K
12. Hua, S.Y., Johnson, D.J.: Research issues on factors influencing the choice of kitting versus line stocking. Int. J. Prod. Res. **48**(3), 779–800 (2010). https://doi.org/10.1080/00207540802456802
13. Limère, V., Van Landeghem, H., Goetschalckx, M., Aghezzaf, E.H., McGinnis, L.F.: Optimising part feeding in the automotive assembly industry: deciding between kitting and line stocking. Int. J. Prod. Res. **50**(15), 4046–4060 (2012). https://doi.org/10.1080/00207543.2011.588625
14. Battini, D., Faccio, M., Persona, A., Sgarbossa, F.: Supermarket warehouses: stocking policies optimization in an assembly-to-order environment. Int. J. Adv. Manuf. Technol. **50**, 775–788 (2010). https://doi.org/10.1007/s00170-010-2555-0
15. Caputo, A.C., Pelagagge, P.M., Salini, P.: A decision model for selecting parts feeding policies in assembly lines. Ind. Manag. Data Syst. **115**(6), 974–1003 (2015). https://doi.org/10.1108/IMDS-02-2015-0054
16. Hanson, R., Medbo, L.: Aspects influencing man-hour efficiency of kit preparation for mixed-model assembly. Procedia CIRP **44**, 353–358 (2016). https://doi.org/10.1016/j.procir.2016.02.064
17. Pegden, C.D., Sturrock, D.T.: Rapid modeling solutions: introduction to simulation and simio. Technical report, Simio LLC (2013)
18. Boysen, N., Emde, S., Hoeck, M., Kauderer, M.: Part logistics in the automotive industry: decision problems, literature review and research agenda. Eur. J. Oper. Res. **242**, 107–120 (2015). https://doi.org/10.1016/j.ejor.2014.09.065

The Rough Interval Shortest Path Problem

Ali Moghanni and Marta Pascoal

Abstract The shortest path problem is one of the most popular network optimization problems and it is of great importance in areas such as transportation, network design or telecommunications. This model deals with determining a minimum weighted path between a pair of nodes of a given network. The deterministic version of the problem can be solved easily, in polynomial time, but sometimes uncertainty or vagueness is encountered. In this work we consider the rough interval shortest path problem, where each arc's weight is represented by a lower approximation interval and an upper approximation interval, which surely contains the real weight value and that may possibly contain the real weight value, respectively. A labeling algorithm is developed to find the set of efficient solutions of the problem.

Keywords Rough sets · Shortest path · Labeling · Efficient solutions

1 Introduction

The shortest path problem is a classical network optimization model with a wide range of applications in areas such as transportation, network design, telecommunications, etc. This model is used whenever we want to minimize a linear function of a path between a certain pair of nodes of a given network, such as the distance, the time or the cost [1].

A. Moghanni · M. Pascoal (✉)
CMUC, Department of Mathematics, University of Coimbra, 3001-501 Coimbra, Portugal
e-mail: marta@mat.uc.pt
URL: https://apps.uc.pt/mypage/faculty/uc25513/

A. Moghanni
e-mail: ali.moghanni@mat.uc.pt
URL: https://sites.google.com/view/alimoghanni/

M. Pascoal
Institute for Systems Engineering and Computers – Coimbra, University of Coimbra, rua Sílvio Lima, Pólo II, 3030-290 Coimbra, Portugal

S. Relvas et al. (eds.), *Operational Research*, Springer Proceedings in Mathematics & Statistics 374, https://doi.org/10.1007/978-3-030-85476-8_5

Traditionally a weight is associated with each directed arc of the network and the goal of the problem is to find the shortest path, with respect to the sum of the weights, from a source to a target. On the other hand, the weights may fluctuate depending on the real problem conditions, like traffic, payload and so on, and this brings the concepts of vagueness and uncertainty. The rough set theory, proposed by Pawlak in 1982 [6, 7] can be seen as a mathematical approach to vagueness. Although the rough set theory overlaps other theories, this approach seems to be of fundamental importance in research areas such as machine learning, intelligent systems, inductive reasoning, pattern recognition, knowledge discovery, decision analysis, and expert systems [3, 8]. In such cases, fuzzy numbers have often been used as the arc weights. However, it is not always easy for a decision maker to determine the membership functions, and it may be more intuitive to express them as rough intervals than as fuzzy numbers [9].

The focus of this work is a shortest path problem between a pair of nodes in a network, the arcs of which are associated with rough intervals. Rough intervals can be compared by means of a partial order relation. Therefore, our purpose is to introduce an algorithm for finding a set of paths said to be efficient for this problem, in the sense that no other path can represent a better alternative. The method results from the combination of the labeling algorithm proposed by Martins [4], modified in order to handle rough intervals, also considering interval arithmetic operations used by Okada and Gen [5].

This paper is structured as follows. In Sect. 2 notation and some preliminary definitions are presented. In Sect. 3, the rough interval shortest path problem (RISPP) is formulated, under the assumption that each arc parameter is defined by a lower approximation and an upper approximation intervals. In Sect. 4, a labeling algorithm is introduced to find efficient paths with respect to the order relation between rough intervals defined before. Section 5 presents computational results, which are followed by concluding remarks.

2 Preliminary Concepts

In this section, we introduce several concepts related with roughness and interval arithmetic used along this paper.

2.1 *Rough Set Theory*

The rough set theory intends to jointly manage vagueness and uncertainty. A basic assumption in this theory is that any concept can be defined through the collection of all the objects that exhibit the properties associated with it. Therefore, a vague concept is assumed to be decomposed in two well-delimited concepts that can be handled independently. Namely, for a vague concept R, it can be formulated:

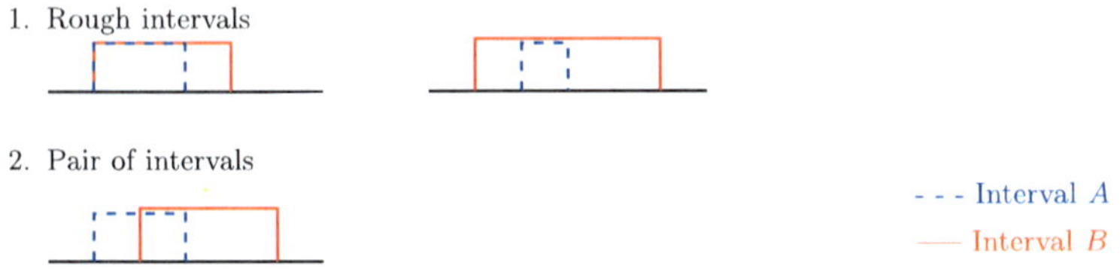

Fig. 1 Rough intervals

1. a lower approximation $\underline{R}$ containing the elements that are surely included in R, and
2. an upper approximation $\overline{R}$ containing those elements that may be included in R, that is, which cannot, beyond doubt, be excluded from R.

Rough intervals are a particular case of rough sets, used to modeling continuous variables. Such an interval is characterized by two parts: a lower approximation interval and an upper approximation interval, which satisfy the rough set lower approximation and upper approximation conditions above. Given a variable x and lower and upper approximation intervals, $\underline{X}$ and $\overline{X}$, these notions can be represented by the following implications:

1. $x \in \underline{X} \Rightarrow x \in X$,
2. $x \notin \overline{X} \Rightarrow x \notin X$.

Then, $\underline{X} \subseteq X \subseteq \overline{X}$. If $\overline{X} - \underline{X} \neq \emptyset$, the interval is said to be rough and is denoted by $(\underline{X}, \overline{X})$. It is said to be crisp, otherwise. It follows from these conditions that a rough interval can be written as an ordered pair $([a_1, a_2], [b_1, b_2])$ such that

$$b_1 \leq a_1 < a_2 \leq b_2.$$

Examples of pairs of intervals that define (the top ones), and that do not define (at the bottom), rough intervals are depicted in Fig. 1.

2.2 *Interval Arithmetic*

In the following some notions on interval arithmetic and interval order relations, used later on to evaluate and to compare paths in a network, are introduced.

Real number intervals can be seen as ordered pairs of real numbers, and thus their arithmetic operations can be defined similarly [2].

Definition 1 Let $A = [a_1, a_2]$ and $B = [b_1, b_2]$ be two intervals of real numbers. Then, the sum $A + B$ is an interval defined as

$$A + B = [a_1 + b_1, a_2 + b_2].$$

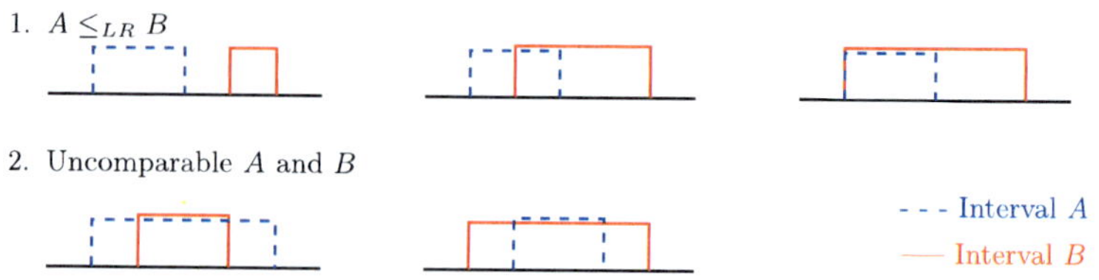

Fig. 2 Comparison between intervals A and B

Example 1 According to this definition, the sum of intervals $[1, 3]$ and $[2, 4]$ is the interval $[1, 3] + [2, 4] = [3, 7]$.

For comparing two intervals, an order relation based on the comparison between the left and right limits is introduced [5].

Definition 2 For any intervals $A = [a_1, a_2]$ and $B = [b_1, b_2]$, the relation $\leq_{LR}$ is defined as

$$A \leq_{LR} B \quad \text{if and only if} \quad a_1 \leq b_1 \text{ and } a_2 \leq b_2.$$

It can be shown that $\leq_{LR}$ is reflexive, anti-symmetric and transitive, and therefore it is an order relation. However it is not a total order relation, as there may be intervals that cannot be compared to one another. This is illustrated in Fig. 2.

3 The Rough Interval Shortest Path Problem

Let $G = (N, A)$ be a directed network with a set of nodes $N = \{1, \ldots, n\}$ and a set of arcs $A \subseteq N \times N$. Each arc $(i, j) \in A$ is associated with a rough interval that can represent the distance between the nodes i and j, or the time for traversing the arc (i, j). This rough interval is denoted by $C_{ij} = (\underline{C}_{ij}, \overline{C}_{ij})$, where $\underline{C}_{ij}$ and $\overline{C}_{ij}$ stand for its lower and upper approximations.

A path in G from node $i_0 \in N$ to node $i_l \in N$ is a sequence

$$\langle (i_0, i_1), (i_1, i_2), \ldots, (i_{l-1}, i_l) \rangle$$

of arcs in A. The RISPP with source node $s \in N$ and target node $t \in N$ can be formulated as follows

$$\begin{aligned} &\text{“minimize”} \sum_{(i,j)\in A} C_{ij} x_{ij} \\ &\text{subject to} \quad \sum_{(i,j)\in A} x_{ij} - \sum_{(j,i)\in A} x_{ji} = \begin{cases} 1 \text{ if } i = s \\ 0 \text{ if } i \in N - \{s, t\} \\ -1 \text{ if } i = t \end{cases} \\ &\qquad x_{ij} \in \{0, 1\}, (i, j) \in A \end{aligned} \tag{1}$$

where $x \in \{0, 1\}^{|A|}$ is the vector with binary values for the components associated with the arcs in the solution. The objective function is the summation of rough intervals, C_{ij}, which is defined as an extension of Definition 1.

Definition 3 Let $(\underline{A}, \overline{A})$ and $(\underline{B}, \overline{B})$ be two rough intervals. Then their sum is defined as follows:

$$(\underline{A}, \overline{A}) + (\underline{B}, \overline{B}) = (\underline{A} + \underline{B}, \overline{A} + \overline{B}).$$

It can be shown that the result of the operation introduced in Definition 3, $(\underline{A}, \overline{A}) + (\underline{B}, \overline{B})$, is still a rough interval.

Lemma 1 *If* $(\underline{A}, \overline{A})$ *and* $(\underline{B}, \overline{B})$ *are two given rough intervals, then*

$$(\underline{A} + \underline{B}, \overline{A} + \overline{B})$$

is also a rough interval.

Additionally in the RISPP different paths need to be compared. In order to do that, the order relation $\leq_{LR}$, given in Definition 2, is now adapted for rough intervals as a dominance relation.

Definition 4 Let $(\underline{A}, \overline{A})$ and $(\underline{B}, \overline{B})$ be two rough intervals. Then, $(\underline{A}, \overline{A})$ dominates $(\underline{B}, \overline{B})$ if and only if

$$\begin{cases} \underline{A} \leq_{LR} \underline{B} \\ \overline{A} \leq_{LR} \overline{B} \end{cases} \text{and } (\underline{A}, \overline{A}) \neq (\underline{B}, \overline{B}).$$

Figure 3 illustrates the dominance relation between several pairs of rough intervals. It also shows that, as expected, not every two rough intervals can be compared, that is, there may be rough intervals $(\underline{A}, \overline{A})$ and $(\underline{B}, \overline{B})$ such that neither the first dominates the latter, nor the opposite holds.

Since not all pairs of rough intervals can be compared, in general there is not a unique solution for the RISPP, but there rather is a set of paths that are not dominated by any other. These solutions are called efficient paths. The purpose of the method presented in the next section is to compute the set of all the efficient paths from s to t in G.

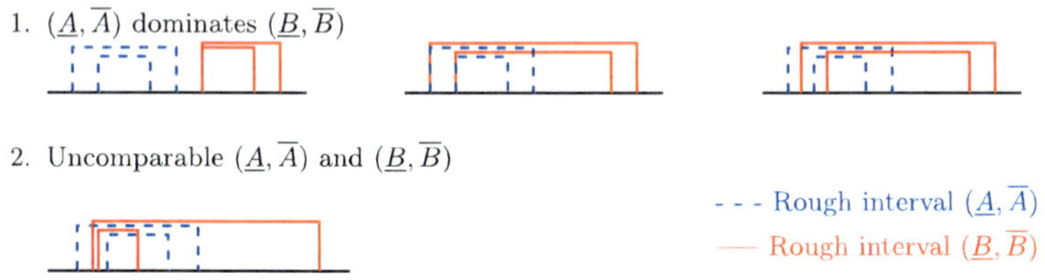

Fig. 3 Comparison between rough intervals $(\underline{A}, \overline{A})$ and $(\underline{B}, \overline{B})$

4 Algorithmic Approach

In this section an algorithm is introduced for finding the set of efficient paths between two nodes with respect to given rough intervals. The algorithm is based on the labeling algorithm proposed by Martins for the multicriteria shortest path problem [4].

As seen before each path from node s to node $i \in N$, p_{si}, has a well defined rough interval $C(p_{si}) = (\underline{C}(p_{si}), \overline{C}(p_{si}))$. Because there may exist more than one path from s to i and because not all of them can be compared, in order to find the efficient rough interval shortest paths from s to t, it is necessary to store information about all the partial paths that are not dominated. Thus each node $i \in N$ is associated with a set of labels, each one representing a path starting in s. Let $i \in N$ be a node of G and let p_{si} represent a path from s to i in G. The q-th label associated with i is the tuple $l_i^q = [C(p_{si}), \pi_i^q, \alpha_i^q]_q$, where:

- $C(p_{si})$ represents the rough interval that gives the lower and the upper approximations for the cost of path p_{si}, and
- $\pi_i^q = j \in N$ is the node that precedes node i in path p_{si}, and
- $\alpha_i^q = k$ is the index of the label of node $j \in N$ in the path that precedes node i, that is, this label is l_j^k.

Two labels representing paths from s to the same node can be compared by looking at the rough interval costs of each one and applying Definition 4. The new definition is presented below.

Definition 5 For some node $i \in N$, we say that the label $[(\underline{C}(p_{si}), \overline{C}(p_{si})), -, -]$ dominates the label $[(\underline{C}(p'_{si}), \overline{C}(p'_{si})), -, -]$ if and only if

$$\begin{cases} \underline{C}(p_{si}) \leq_{LR} \underline{C}(p'_{si}) \\ \overline{C}(p_{si}) \leq_{LR} \overline{C}(p'_{si}) \end{cases} \text{and } (\underline{C}(p_{si}), \overline{C}(p_{si})) \neq (\underline{C}(p'_{si}), \overline{C}(p'_{si})).$$

The method stores node labels as temporary as soon as they are set. These labels may dominate others, which are then deleted. They may also be deleted later themselves, due to the appearance of newer labels that dominate it.

A temporary label becomes permanent (or definite) if it is chosen to be scanned in a suitable order and, thus, to extend the partial paths. This goal can be accomplished by selecting the temporary labels by lexicographic order. The lexicographic order for rough intervals results from the notion of lexicographic order for 4-tuples.

Definition 6 Let $A = ([a_1, a_2], [a_3, a_4])$ and $B = ([b_1, b_2], [b_3, b_4])$ be two rough intervals, such that $a_3 \leq a_1 < a_2 \leq a_4$ and $b_3 \leq b_1 < b_2 \leq b_4$. We say that A is lexicographically smaller than or equal to B, denoted by $A \leq_{Lex} B$, if and only if

$$A = B$$

or if there exists $i \in \{1, 2, 3, 4\}$ such that

$$a_i < b_i \text{ and } a_j = b_j, \text{ for all } j < i.$$

It is remarked that the lexicographic order is a total order relation, which ensures that a minimum label can be found at any iteration of the algorithm.

Definition 7 Given the node $i \in N$, the label $[(\underline{C}(p_{si}), \overline{C}(p_{si})), -, -]$ is lexicographically smaller than the label $[(\underline{C}(p'_{si}), \overline{C}(p'_{si})), -, -]$ if and only if

$$(\underline{C}(p_{si}), \overline{C}(p_{si})) \leq_{Lex} (\underline{C}(p'_{si}), \overline{C}(p'_{si})).$$

From a permanent label of some node $i \in N$, a temporary label is assigned to every node $j \in N$, for any arc $(i, j) \in A$, provided that it is not dominated by any other. The labeling method for finding the set of efficient rough interval shortest paths is outlined in Algorithm 1. It should be noted that changing node t in line 12 of Algorithm 1 to any other node $j \in N$ also allows to obtain the rough interval shortest path from s to j.

Algorithm 1: Labeling algorithm for the rough interval shortest path problem

Input: A graph G with one rough interval associated with each arc, a source node s, and a target node t
Output: Rough interval shortest path from s to t

1 Assign the temporary label $l_s^1 = [([0, 0], [0, 0]), -, -]_1$ to node s
2 **while** *the set of temporary labels is not empty* **do**
3   Find the lexicographically smallest temporary label. Let it be the label l_i^q
4   Set this label as permanent
5   **for** $j \in N$ *such that* $(i, j) \in A$ **do**
6         $k \leftarrow$ current number of labels of node j
7         $(\underline{C}(p_{sj}), \overline{C}(p_{sj})) \leftarrow (\underline{C}(p_{si}) + \underline{C}_{ij}, \overline{C}(p_{si}) + \overline{C}_{ij})$
8         $\pi_j^{k+1} \leftarrow i$
9         $\alpha_j^{k+1} \leftarrow q$
10   Assign the temporary label $l_j^{k+1} = [(\underline{C}(p_{sj}), \overline{C}(p_{sj})), \pi_j^{k+1}, \alpha_j^{k+1}]_{k+1}$ to node j
11   Delete the temporary labels of node j that correspond to paths from s to j dominated by the new path

12 **for** *any efficient path from s to t, p_{st}* **do**
13     $i \leftarrow t$
14     $q \leftarrow$ index of the label corresponding to path p_{st}
15   **while** $i \neq s$ **do**
16   Output node i
17         $i \leftarrow \pi_i^q$
18         $q \leftarrow \alpha_i^q$
19   Output node s

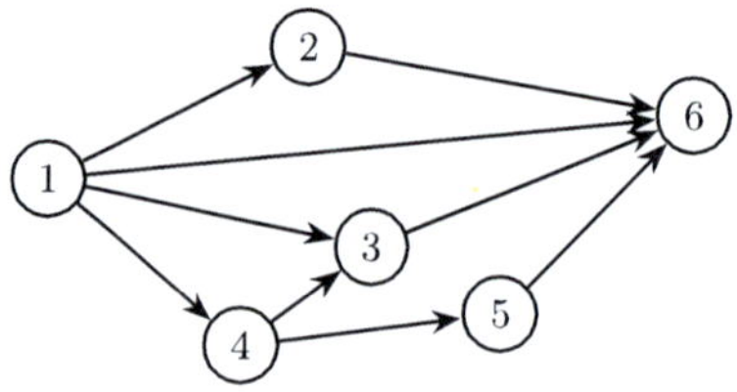

(i,j)	$(\underline{C}_{ij}, \overline{C}_{ij})$
$(1,2)$	$([16,29],[10,30])$
$(1,3)$	$([22,48],[8,69])$
$(1,4)$	$([13,23],[6,36])$
$(1,6)$	$([33,60],[20,81])$
$(2,6)$	$([15,30],[8,50])$
$(3,6)$	$([18,25],[4,47])$
$(4,3)$	$([7,17],[1,20])$
$(4,5)$	$([12,27],[9,30])$
$(5,6)$	$([5,8],[2,13])$

Fig. 4 Network $G = (N, A)$ and rough interval arc weights

To illustrate the Algorithm 1 we apply it to the network depicted in Fig. 4. The source and the target nodes are $s = 1$ and $t = 6$, respectively. The arc rough intervals are shown in the table on the right-hand side of Fig. 4.

The method starts by assigning the temporary label $l_1^1 = [([0,0],[0,0]), -, -]_1$ to node $s = 1$. This label is set as a permanent label after the first iteration of the **while** loop at line 2. When scanning node $s = 1$, the nodes 2, 3, 4 and 6 are labeled with

- $l_2^1 = [([16,29],[10,30]), 1, 1]_1$,
- $l_3^1 = [([22,48],[8,69]), 1, 1]_1$,
- $l_4^1 = [([13,23],[6,36]), 1, 1]_1$,
- $l_6^1 = [([33,60],[20,81]), 1, 1]_1$,

respectively. In the next iteration of the loop, the label l_4^1 is the lexicographically smallest temporary label. When scanning it the node 5 is also labeled, with $l_5^1 = [([25,50],[15,66]), 4, 1]_1$ and a new label of node 3 is created, $l_3^2 = [([20,40],[7,56]), 4, 1]_2$ that dominates the former, which is thus deleted. Then l_2^1 is the lexicographically smallest temporary label, thus the node 6 is labeled with $l_6^2 = [([31,59],[18,80]), 2, 1]_2$, which dominates the former label l_6^1. In the next step of the algorithm, the label l_3^2 is the lexicographically smallest temporary label, thus node 6 is labeled again, this time with $l_6^3 = [([38,65],[11,103]), 3, 2]_3$, which neither dominates the former labels, nor the opposite holds.

The procedure continues until there are no temporary labels left to scan. At the end two efficient paths are obtained from $s = 1$ to $t = 6$:

- $p_1 = \langle (1,4), (4,5), (5,6) \rangle$,
- $p_2 = \langle (1,4), (4,3), (3,6) \rangle$.

The final results for each node, that is, the labels of efficient paths from s to any node in the network are reported in Table 1.

Table 1 Labels of the efficient paths from $s = 1$ to any node in the network in Fig. 2

Node	Labels
1	$[([0, 0], [0, 0]), -, -]_1$
2	$[([16, 29], [10, 30]), 1, 1]_1$
3	$[([20, 40], [7, 56]), 4, 1]_2$
4	$[([13, 23], [6, 36]), 1, 1]_1$
5	$[([25, 50], [15, 66]), 4, 1]_1$
6	$[([30, 58], [17, 79]), 5, 1]_4$
	$[([38, 65], [11, 103]), 3, 2]_3$

5 Computational Experiments

This section is dedicated to assessing the empirical performance of Algorithm 1 for two sets of instances of the RISPPs. The algorithm was implemented in Matlab (R2018a). The tests ran on a 64-bit PC with an Intel® Core™ i7-7500U at 3.5 GHz with 12 GB of RAM.

A first set of tests was performed on randomly generated networks with $n = 1\,000, 5\,000, 10\,000$ nodes, $m = dn$ arcs, for average degrees $d = 5, 10$. The arc rough intervals are defined by 4 integer numbers, generated uniformly in $[1, 1\,000]$. These instances are denoted by $R_{n,d}$.

The second set of instances are grid networks, denoted by G, where nodes are arranged in a rectangular, or square, grid with given height and width. Any pair of adjacent nodes is connected by an arc in both directions, and each arc is associated with a rough interval generated as in the random networks case. The characteristic of this set of networks are summarized in Table 3.

In order to analyze the performance of Algorithm 1, the code ran over 30 different instances for each dimension, and the mean CPU times and number of efficient paths were calculated.

For the two cases, the mean values increase rapidly with the size of the network, given by n and d.

Table 2 and Fig. 5 show the mean CPU times and the mean number of efficient paths obtained by Algorithm 1 for the first set of experiments. Those results show

Table 2 Mean results for Algorithm 1 in random networks

$R_{n,d}$	CPU time (s)	# Efficient paths
$R_{1\,000,5}$	0.925	3.6
$R_{1\,000,10}$	1.685	4.0
$R_{5\,000,5}$	5.860	3.5
$R_{5\,000,10}$	19.853	5.3
$R_{10\,000,5}$	29.257	4.9
$R_{10\,000,10}$	82.408	5.3

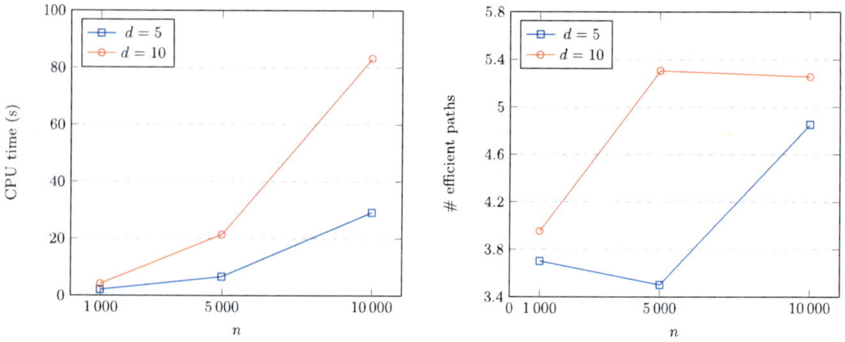

Fig. 5 Mean results for Algorithm 1 in random networks

Table 3 Mean results for Algorithm 1 in grid networks

Name	Size	n	m	CPU time (s)	# Efficient paths
G_1	2×50	100	296	0.647	23.8
G_2	50×2	100	296	0.808	43.1
G_3	10×10	100	360	3.408	504.1
G_4	2×72	144	428	1.703	61.2
G_5	72×2	144	428	1.520	65.9
G_6	3×48	144	474	2.706	84.9
G_7	48×3	144	474	2.370	123.3
G_8	4×36	144	496	3.325	250.3
G_9	36×4	144	496	2.936	257.3
G_{10}	12×12	144	528	37.605	1345.9
G_{11}	3×75	225	744	29.738	268.9
G_{12}	75×3	225	744	38.367	232.0
G_{13}	5×45	225	800	57.414	970.1
G_{14}	45×5	225	800	84.661	858.1

that the CPU times increase rapidly, up to 83 s, with the increase of n and d. The plots also show a fairly similar growth of the number of efficient paths with the number of nodes of the network.

The mean CPU times and number of efficient paths of Algorithm 1 when applied to the grid networks are reported in Table 3 and in Fig. 6. It is remarked that the problem is much more difficult for square grids than for rectangular grids with similar number of nodes, which is a consequence of the higher number of efficient paths on this type of grids. All instances were solved in less than 90 s, but the RISPP was also more difficult to solve for grid instances than for random instances. The mean number of computed efficient paths ranged from 23.8 to 1345.9.

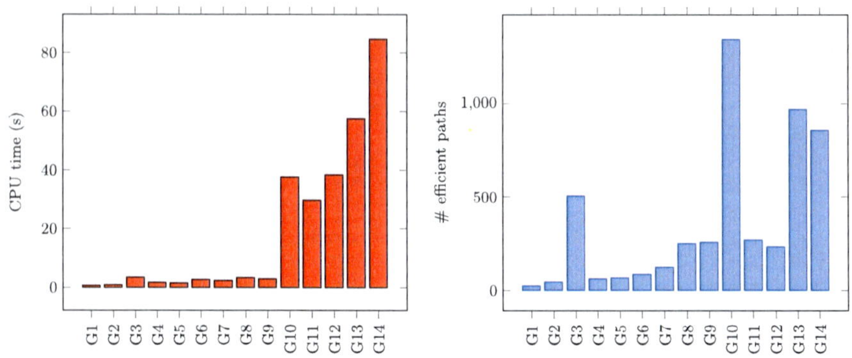

Fig. 6 Mean results for Algorithm 1 in grid networks

6 Conclusions

This paper addressed the rough interval shortest path problem. It was assumed that the arc weights are defined as pairs of intervals, a lower approximation interval, which surely contains the arc weight, and an upper approximation interval, which may contain the arc weight. A labeling method was presented which is able to find the efficient paths from a single source to all other nodes with respect to the rough interval arc weights. The proposed algorithm was tested regarding the CPU time and the number of produced solutions over random and grid networks. The obtained results are promising as the algorithm was always able to obtain a set of efficient rough interval shortest paths within less than 2 min, in average. Future lines of research may include studying a bicriteria version of the problem with rough interval costs and capacities.

Acknowledgements This work was partially supported by the Portuguese Foundation for Science and Technology (FCT) under project grants UID/MAT/00324/2019 and UID/MULTI/00308/2019. The work was also partially financially supported by project P2020 SAICTPAC/0011/2015, co-financed by COMPETE 2020, Portugal 2020 - Operational Program for Competitiveness and Internationalization (POCI), European Union's European Regional Development Fund, and FCT, and by FEDER Funds and National Funds under project CENTRO-01-0145-FEDER-029312.

References

1. Ahuja, R., Magnanti, T., Orlin, J.: Network Flows: Theory, Algorithms and Applications. Prentice Hall, Englewood Cliffs (1993)
2. Alefeld, G., Herzberger, J.: Introduction to Interval Computations. Computer Science and Applied Mathematics, 1st edn. Elsevier Science; Academic, Cambridge, (1983)
3. Mahajan, P., Kandwal, R., Vijay, R.: Rough set approach in machine learning: a review. Int. J. Comput. Appl. **56**, 1–13 (2012)
4. Martins, E.: On a multicriteria shortest path problem. Eur. J. Oper. Res. **16**(2), 236–245 (1984)

5. Okada, S., Gen, M.: Order relation between intervals and its application to shortest path problem. Comput. Ind. Eng. **25** (1993)
6. Pawlak, Z.: Rough sets. Int. J. Comput. Inf. Sci. **11**, 341–356 (1982)
7. Pawlak, Z.: Imprecise Categories, Approximations and Rough Sets, pp. 9–32. Springer, Netherlands (1991)
8. Rebolledo, M.: Rough intervals-enhancing intervals for qualitative modeling of technical systems. Artif. Intell. **170**, 667–685 (2006)
9. Slowinski, R., Vanderpooten, D.: A generalized definition of rough approximations based on similarity. IEEE Trans. Knowl. Data Eng. **12**, 331–336 (2000). Mar

Reinforcement Learning for Robust Optimization: An Application in Kidney Exchange Programs

Tiago Monteiro, João Pedro Pedroso, Ana Viana, and Xenia Klimentova

Abstract Kidney Exchange Programs allow an incompatible patient-donor pair, whose donor cannot provide a kidney to the respective patient, to have a transplant exchange with another pair in a similar situation. The associated combinatorial problem of finding such exchanges can be represented by a graph: nodes represent incompatible pairs and arcs represent compatibility between donor in one pair and patient in the other. This problem has some uncertainty, which in the literature has been commonly addressed in the following ways: expected utility, fall-back mechanisms and robust optimization. We propose an alternative interactive tool to support decision makers (DMs) on choosing a solution, taking into account that some pairs may become unavailable. For a given solution the predicted performance is evaluated under multiple scenarios generated by Monte Carlo Tree Search (MCTS). The root node of the tree corresponds to no failures. From there, a tree of failure possibilities is generated, each of them corresponding to a different scenario. A solution is determined for every particular scenario. At the end, each solution is evaluated under each scenario. Scenarios are grouped based on the cardinality of the set of failing vertices, and average results for each cardinality are considered. Finally, Pareto dominated solutions are filtered out and the non-dominated average solutions are displayed and compared with the worst case scenario. The tool visually drives DMs in the process of choosing the best solution for their particular preferences.

T. Monteiro (✉) · J. P. Pedroso · A. Viana · X. Klimentova
INESC TEC, Campus da FEUP, 4200-465 Porto, Portugal
e-mail: tiago.p.monteiro@inesctec.pt

J. P. Pedroso
e-mail: jpp@fc.up.pt

A. Viana
e-mail: ana.viana@inesctec.pt

X. Klimentova
e-mail: xenia.klimentova@inesctec.pt

J. P. Pedroso
Faculdade de Ciências, Universidade do Porto, 4169-007 Porto, Portugal

A. Viana
ISEP - School of Enginnering, Polytechnic of Porto, 4200-072 Porto, Portugal

S. Relvas et al. (eds.), *Operational Research*, Springer Proceedings in Mathematics & Statistics 374, https://doi.org/10.1007/978-3-030-85476-8_6

Keywords Kidney exchange programs · Data uncertainty · Monte Carlo Tree Search · Interactive decision making

1 Introduction

Currently, the most effective treatment for patients that suffer from chronic kidney disease is transplantation. But finding suitable kidneys for transplantation can be difficult: they may come from a deceased donor (though the supply is relatively low when compared with the demand, resulting in long waiting lists), or from living donors (where the major barrier is the fact that the willing donor is often blood and/or tissue type incompatible with the intended patient) [5, 12]. That circumstance led to the organization of kidney exchange programs, which allow incompatible patient and living donor pairs, whose donor cannot provide a kidney to the respective patient, to have an "exchange" with another pair in a similar situation [10].

The simplest case of an exchange, illustrated in Fig. 1 (left), involves only two incompatible patient-donor pairs, (P_1, D_1) and (P_2, D_2). The donor from the first pair (D_1) gives a kidney to the patient of the second pair (P_2), and the patient of the first pair (P_1) gets a kidney from the donor of the second pair (D_2). When k incompatible pairs are involved in an exchange, this concept extends to k-exchange, with $k \geq 2$. Figure 1 (right) illustrates an exchange between three incompatible pairs. Donor D_1 gives a kidney to patient P_2, donor D_2 to patient P_3 and donor D_3 to patient P_1. However, a bound on the number of pairs involved in an exchange is necessary for logistic reasons, and also to reduce the number of affected patients when last-minute donor resignation occurs or new incompatibilities are detected [10].

The optimization problem underlying kidney exchange programs can be formulated as an Integer Program (IP) that aims at maximizing the number of transplants. For $k \geq 3$ and bounded the problem is NP-hard [10].

Initial approaches considered that all data associated to the problem was certain, but many recent studies consider data uncertainty, as in practice some of planned transplants may not occur. One reason for this derives from the fact that new incompatibilities between donors and patients selected for transplant may be detected at the last-minute, preventing the operation from proceeding [2]. Other transplantation

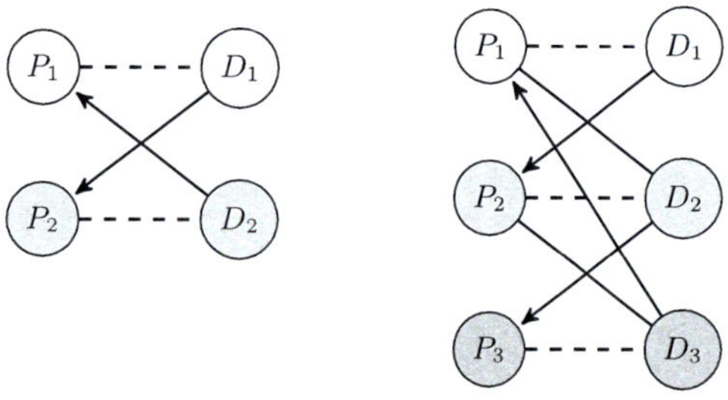

Fig. 1 An exchange between two (left) and three (right) incompatible pairs. Solid lines represent preliminary assessment compatibility and arrows define a possible exchange [10]

failure reasons are: the patient receives a kidney from the deceased donor list; illness changes a patient or donor's antigen incompatibilities; patient or donor become too ill for surgery [7].

In the literature different methods were proposed for handling uncertainty: expected utility, fall-back mechanisms and robust optimization. The former considers probability of failure of vertices and/or arcs and computes the expected size of a cycle rather than the actual one [5, 10, 14]. The second consists in considering fall-back options offered by all sub-cycles and sub-chains within the selected cycles and chains (e.g., give priority to 3-exchanges that can result in 2-exchanges if a pair fails) [3, 12, 13]. In the latter, the optimal solution for the worst case scenario is found [9].

In this work, we propose an alternative interactive tool that visually drives DMs in the process of choosing the best solution for their particular preferences by showing the predicted performance in cases (scenarios) where some pairs become unavailable. For that, we use a scenario-based approach to robust optimization that aims at maximizing the number of exchanges in the initial and final solution given a set of scenarios. The generation of all possible combinations for pairs that become unavailable to be used as scenarios is unpractical as it requires a high computational effort. Instead of that, we use a Reinforcement Learning method, Monte Carlo Tree Search [4, 19], where the aim is to focus on the most damaging scenarios. In the proposed procedure, the root node of the tree corresponds to an optimal solution under the scenario of no failures. Then a tree of failure possibilities is generated, where each node corresponds to a particular set of failing pairs, defining a scenario. For each scenario we determine the corresponding optimal solution. At the end, each solution is evaluated under each of the scenarios considered, with scenarios grouped by the associated cardinality of set of failing vertices. Average results are obtained for each cardinality value. Finally, Pareto dominated solutions are filtered out for each number of failing vertices, and the non-dominated solutions are displayed and compared with the worst case scenario.

2 Literature Review

The concept of kidney exchange program for incompatible patient-donor pairs was introduced in 1986 by Rapaport as an alternative to deceased donor programs [16]. Since then, due to its importance, several studies were presented in the literature for the kidney exchange problem and its variants.

An optimal solution for this problem can be obtained in polynomial time with the Edmonds' algorithm when only 2-exchanges are considered [8]. If we consider k-exchanges and k is unbounded, the problem turns into an assignment problem and can also be solved in polynomial time. It is shown in [18] that if we only consider 2-exchanges, an important number of transplants is missed, being $k = 3$ the number that captures the most realizable pairings. Improvements are not significant if we consider larger values for k. However, for $k \geq 3$ and bounded the problem is known to be NP-hard and difficult to solve efficiently for large instances [1]. Integer programming is the natural framework for modeling and solving a KEP. In [6] a set of IP models is presented, analyzed and compared.

Most recent models for this problem consider the possibility of a planned transplant not occurring. Failures, the term usually applied for that context, can happen for several reasons, such as sickness, pregnancy or death of the candidate or donor, scheduling aspects, or if a crossmatch test contradicts the presumed compatibility of the virtual crossmatch.

In the literature there are different methods to handle failure. In [14] the author proposes an approach considering probabilities for pairs' failure and/or new incompatibilities and maximizes the expected number of transplants. The authors of [10] propose a new tree search algorithm for calculating the expected number of transplants in a KEP. It extends the work in [14] by allowing greater values for k. Besides that, it also proposes a new scheme for rearrangement of vertices in cycles with failure. The authors in [5] also propose a model for maximizing the expected utility when arcs fail. They present a simulation system which assists clinicians in the evaluation of different kidney allocation strategies. In [12] it is taken into consideration the probability that a predicted compatibility will result in an actual transplant operation. It allows one or more contingency allocations in case the original planned exchanges fail. The work in [2] proposes a method for partitioning the original graph in many smaller subsets, referred as locally relevant sub-graphs (LRS). After that, methods that calculate the expected utility of transplant rearrangements are proposed: exact calculation, inclusion-exclusion, matrix formulation and Monte Carlo sampling (estimation). These approaches also consider the fallback option. In [13] and [3] it is given priority to 3-exchanges that result in 2-exchanges if a pair fails. Finally in [9] authors found the optimal solution for the worst case scenario considering a given number of failures.

3 Problem Description

In this section we describe the Integer Programming formulation used in our work to obtain the optimal solution associated to each scenario and describe the MCTS procedure used for scenario generation.

3.1 *Integer Programming Model—Cycle Formulation*

Let $G(V, A)$ be a directed graph with the set of vertices $V = 1, \ldots, |V|$ consisting of all incompatible patient-donor pairs and the set of arcs A designating compatibilities between the vertices. Two vertices $i, j \in V$ are connected by arc (i, j) if the patient in pair j is compatible with the donor in pair i. To each arc $(i, j) \in A$ is associated a weight w_{ij}. If the objective is to maximize the total number of transplants, then $w_{ij} = 1, \forall (i, j) \in A$.

The kidney exchange problem can be defined as follows: find a packing of vertex-disjoint cycles with length at most k having maximum weight. This can be modelled by the well-known cycle formulation [6]. Let $\mathcal{C}(k)$ be the set of all cycles in G with length at most k. Define a variable z_c for each cycle $c \in \mathcal{C}(k)$ such that:

$$z_c = \begin{cases} 1 & \text{if cycle } c \text{ is selected for the exchange,} \\ 0 & \text{otherwise.} \end{cases}$$

Denote by $V(c) \subseteq V$ the set of vertices which belong to cycle c, and by $w_c = \sum_{(i,j)\in c} w_{ij}$ the weight of the cycle. The model can now be written as follows:

$$\text{Maximize} \quad \sum_{c\in\mathcal{C}(k)} w_c z_c \tag{1a}$$

$$\text{Subject to:} \quad \sum_{c:i\in V(c)} z_c \leq 1 \qquad \forall i \in V \tag{1b}$$

$$z_c \in \{0, 1\} \qquad \forall c \in \mathcal{C}(k). \tag{1c}$$

The objective function (1a) maximizes the weighted number of transplants and constraints (1b) ensure that each vertex is in at most one of the selected cycles (i.e., each donor may donate, and each patient may receive only one kidney).

3.2 *Monte Carlo Tree Search*

Monte Carlo Tree Search is a method used for exploration of a search tree and exploitation of its most promising regions. This algorithm is based on Monte Carlo simulation, where the reward associated with each node is estimated using the result of random simulations started from that node. Each iteration of MCTS can be divided into the following four steps [15]:

1. **Selection:** starting from the root node we select the most "promising" child node (the one with highest associated utility). For determination of the utility value, we use the Upper Confidence Bound for Trees (UCT) algorithm [11] that aims at reaching a balance between the exploration and exploitation of the search tree space. The utility $U(n)$ of node n is calculated based on the following formula:

$$U(n) = X(n) + E(n) \tag{2}$$

   where $X(n)$ and $E(n)$ are, respectively, the exploitation and exploration utility associated with n. At each node, the child with maximum $U(n)$ is selected, until an unexpanded node is reached [15].
   $X(n)$ is a $[0, 1]$ normalized value associated with the best and worst solutions obtained up to the current iteration. The exploration component is obtained through the following formula:

$$E(n) = c\sqrt{\frac{\ln s_{p(n)}}{s_n}} \tag{3}$$

where c is an exploration parameter, normally equal to $\sqrt{2}$, $s_{p(n)}$ is the number of simulations done under parent node $p(n)$ and s_n is the number of simulations done under child node n.

2. **Expansion:** one or more children nodes of the selected node n are created by applying possible actions. We use single expansion, in which only one child node is created. But there are more options, like full expansion, where all children nodes are created at the same time.
3. **Simulation:** nodes created in the expansion step perform a simulation until they reach a terminal state, and the obtained value is recorded. In our case, simulation consists of turning pairs unavailable to be included in cycles, in a random way, until no more cycles can be formed (and we reach a leaf node). The value associated to the leaf node corresponds to the number of nodes visited until the leaf is reached (i.e., its depth).
4. **Backpropagation:** propagates the result obtained in the current simulation step up the tree, through nodes that had been selected, and update their statistics.

Although MCTS is usually used to obtain a solution, in our case we are more interested in the generation of child nodes to be used as scenarios with a specific amount of pairs failing. These scenarios allow us to proceed with simulation tests with the objective of predicting the behavior of each solution (obtained for each scenario), grouped by the amount of pairs failing. The next section describes the procedure in more detail.

3.3 Algorithm

The procedure used to obtain the best solutions for each set of failing pairs (i.e., for each scenario) can be described by the following steps (see also Algorithm 1):

1. After all initializations are made, the first step is to use the IP model to determine an optimal solution for the instance considered;
2. The optimal solution is used as root node for the MCTS execution. Starting from the root node, a tree of failure possibilities is generated; each child of a node is a scenario where one additional pair of the parent's solution becomes unavailable. The execution ends when a predefined number of iterations is reached.
3. After MCTS execution ends it outputs a set of scenarios; since paths where failures quickly lead to no cycles are reinforced, this set is expected to correspond to worst-case situations. Scenarios are then grouped by the number of associated failures (one group for scenarios with a single failure, another one for scenarios with two failures, etc.) and, for each scenario created, the cycle formulation is used to obtain an optimal solution, respecting the constraint that failing pairs

characterizing such scenario are not allowed to participate in that solution; we filter out the worst solutions, i.e., solutions for which the optimal value is lower than $r\%$ (in our tests, 95%) of the optimal solution (without failures). These solutions are discarded for the next steps.

4. We are now able to assess each solution under each of the scenarios. This assessment is done progressively, for scenarios with an increasing number of failures. Hence, a solution is firstly evaluated under scenarios that consider one failure, and the average result is saved; then, it is evaluated under scenarios with two failures, and the average value is retained; and so on, until reaching the maximum number of failures generated by MCTS.
5. After all the solutions are evaluated in all scenarios, we filter out the Pareto dominated solutions, and display only the non-dominated ones.

Algorithm 1: Reinforcement Learning for Robust Optimization: KEP

Input:
- instance graph $G = (V, A)$
- max cycles length k
- number of iterations N for MCTS
- minimum quality r for solution acceptance

Output: Pareto non-dominated solutions graph.

$x^* \leftarrow \mathrm{opt}(G, k)$
$\mathcal{X} \leftarrow \{x^*\}$
$C_i = \emptyset \quad$ for $i \in \{1, \dots, |V|\}$
for $S \in MCTS(G, N, x^*)$ **do**
    $i \leftarrow |S|$ // number of failures in scenario S
    $C_i \leftarrow C_i \cup S$ // add S to set C_i of scenarios with cardinality i
    $G' \leftarrow (V \setminus S, \{(i, j) \in A : i, j \notin V \setminus S\})$
    $x' \leftarrow \mathrm{opt}(G', k)$
    if $\mathrm{obj}(G', x') \geq r\, \mathrm{obj}(G, x^*)$ **then**
        $\mathcal{X} \leftarrow \mathcal{X} \cup \{x'\}$
$M = \max\{i \in 1, \dots, |V| : C_i \neq \emptyset\}$ // maximum number of failures in the MCTS
$\mathcal{P}_i = \emptyset \quad$ for $i \in \{1, \dots, M\}$
for $j \in \{1, \dots, |\mathcal{X}|\}$ **do**
    $x' \leftarrow \mathcal{X}[j]$
    for $i \in \{1, \dots, M\}$ **do**
        $f_{ji} \leftarrow 0$
        for $S \in C_i$ **do**
            $G' \leftarrow (V \setminus S, \{(i, j) \in A : i, j \notin V \setminus S\})$
            $f_{ji} \leftarrow f_{ji} + \mathrm{obj}(G', x')$
        $f_{ji} \leftarrow f_{ji}/|C_i|$
        if $\nexists j' \in \mathcal{P}_i : f_{j'k} \geq f_{jk}, \forall k \leq i$ *and* $f_{j'k} > f_{jk}$ *for some* $k \leq i$ **then**
            $\mathcal{P}_i \leftarrow \mathcal{P}_i \cup \{j\}$ // no point dominates j, add it to the Pareto index set
            $\mathcal{P}_i \leftarrow \mathcal{P}_i \setminus \{j' \in \mathcal{P}_i : f_{jk} \geq f_{j'k}, \forall k \leq i$ and $f_{jk} > f_{j'k}$ for some $k \leq i\}$
return $\{\mathcal{X}[j] : j \in \mathcal{P}_i\}$ for $i \in \{1, \dots, M\}$

In Algorithm 1 we formalize the description of these steps. Step 1 is performed in line 1, where $\text{opt}(G, k)$ function solves the cycle formulation (1a)–(1c) for instance G and returns an optimal solution x^* that is added to the set of solutions $\mathcal{X}$ in line 2. After, as described in steps 2 and 3, $MCTS(G, N, x^*)$ is executed for a predefined number of iterations, returning a set of scenarios. Scenarios are aggregated in sets according to the associated number of failures (line 5–6). For each an optimal solution is obtained in lines 7–8. If the solution objective value is higher than $r\%$ of the optimal solution without failures (in our tests, r was set to 95%), it is added to the set of solutions $\mathcal{X}$ (lines 9–10). Step 4 is implemented in lines 11–19. Here, each solution x' is assessed under each set of scenarios $\mathcal{C}_i$ and its average value under each set is calculated. Finally, lines 20–23 process the results to display: for each number of vertices failing ($i = 1,\ldots, M$) we calculate $\mathcal{P}_i$ as the set of solutions that are not dominated for up to i. The information stored in those sets will be used whenever we want to display results for up to i failures.

4 Results

For the purpose of illustration of the method we propose, in this section we present results for one test instance with 50 pairs,[1] $k = 3$ and using a number of iterations $N = 10000$ in Algorithm 1. We used Python (version 2.7), Gurobi as the optimizer for solving optimization problem (1a)–(1c) and Plotly's python library for plotting the interactive graphs. As basis of this work we used code developed in Ph.D. thesis [17].

Our methodology is a decision support tool that predicts the behavior of solutions in case some pairs become unavailable and helps the DMs to choose the solution that best fits their interests. For this particular case, providing a single solution to the DMs would be unrealistic and hide potentially relevant information. Therefore, we provide average results for Pareto non-dominated solutions in each set of scenarios, through an interactive software solution. The graphical user interface makes the solution behavior comprehension much easier (each solution is associated a different color and the worst cases are represented by dashed lines) and allows actions such as: displaying the average number of transplants for each number of pairs failing, up to a selected number of failures; compare solutions obtained for different parameterizations of the number of failures and/or the respective worst case scenario.

In Fig. 2 it is possible to observe the output at an initial stage (where all the solutions are displayed). Points represent the average number of transplants (values f_{ij} in Algorithm 1) of pre-selected solutions (step 5 of the algorithm) for different values on the number of failing vertices (horizontal axis). The number of scenarios generated for each number of pairs that become unavailable can be seen in the line below the horizontal axis (*Scenarios: 0-n_0, 1-n_1, ...,* where n_i is the number of scenarios considered for i failures). In the right-hand side, there is a list of labels

[1] Available in: http://www.dcc.fc.up.pt/~jpp/code/KEP/small/.

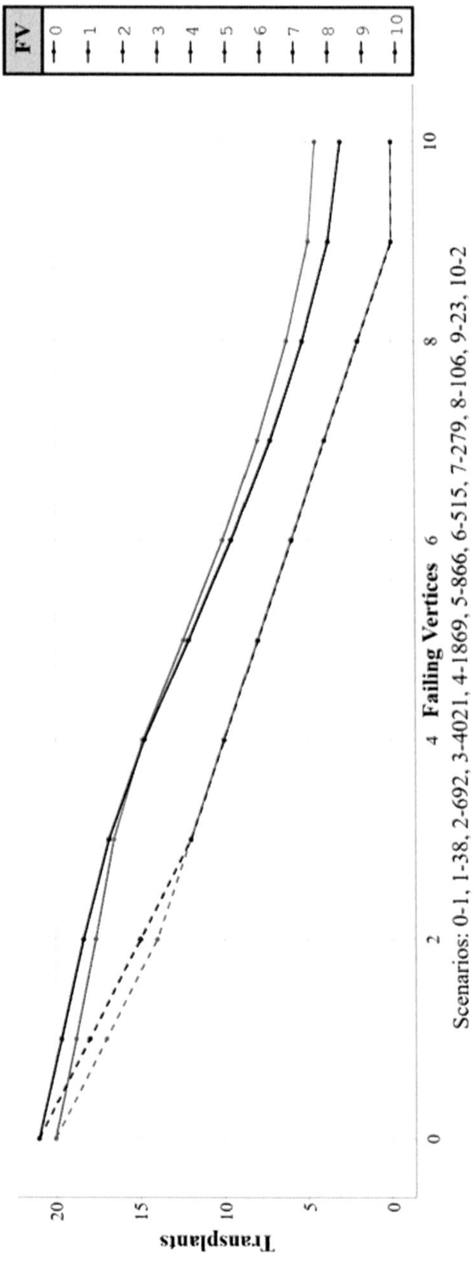

Fig. 2 Output obtained for an instance with 50 pairs, $k = 3$ and 10000 MCTS iterations

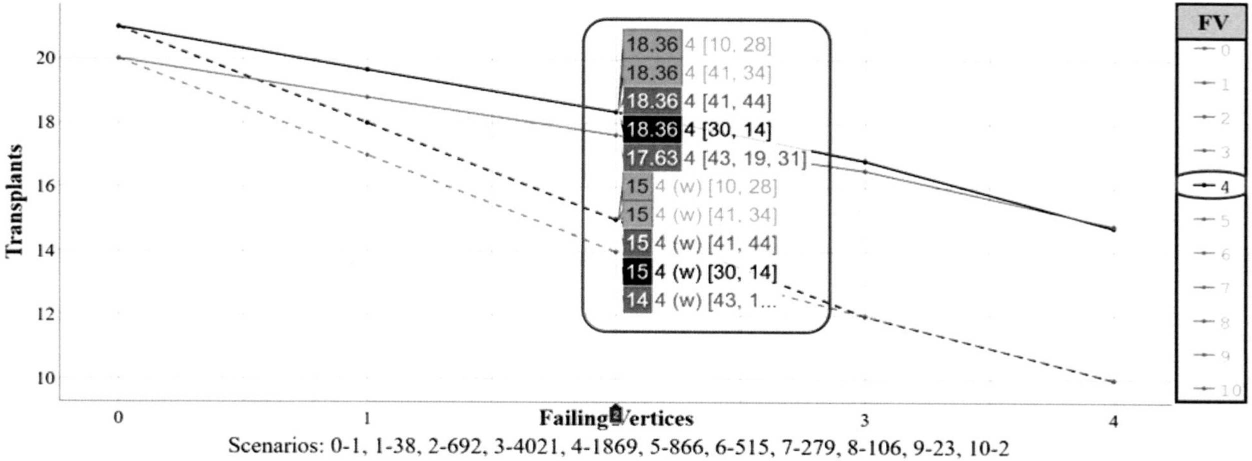

Fig. 3 Interactivity of the software application

(titled FV) that allows DMs to select (or deselect) the number of failures (from 0 to 10) for which they want solutions to be displayed.

For the instance depicted in Fig. 2 we can see that the optimal solution for no failures is composed of 21 pairs. There are multiple alternative solutions with the same value, though they cannot be seen in the figure due to overlap. The multiplicity also holds for solutions involving only 20 pairs. Solutions involving less pairs are not displayed as they do not reach the lower bound established ($0.95 \cdot 21 = 19.95$, see step 3 Sect. 3.3). For this particular case, by setting the limit for MCTS execution to 10000 iterations, scenarios with up to 10 failures are generated. The possibility of the solutions behavior not being uniform can be observed in this case: the solutions that have best results for scenarios with more than four pairs failing are not as good as the results obtained for scenarios where at most three pairs fail.

Figure 3 exemplifies the interactivity of the software. Starting from the same graphic present in Fig. 2, label "4" is selected (in the right-hand side list "FV"), while all the others are deselected (i.e., only non dominated solutions for scenarios with up to four failures are displayed). The solutions average behavior in simulations involving scenarios of up to four failures and the respective worst case (distinguished by using dashed lines and with the label "(w)" in the *hover legend*) are displayed. That solution also provides the average value obtained for each solution (assessed under all scenarios of two failures), the indication on the vertices that failed in the scenario that originated that solution (in between square brackets) and the minimum number of transplants of that solution under all 2-failure scenarios.

It is unquestionable that, with the usage of our tool, decision making is supported by relevant information that is not available when only one solution is proposed. One limitation observed in our methodology is that the amount of scenarios used may not be proportional to the number of possible combinations of pairs failing. This happens because MCTS primarily exploits the most "promising" parts of the tree search space, where failures damage the current solution the most. In contrast, the importance of simulations under scenarios with a higher number of failures can be small, because in practice, it may be rare that scenarios with these worst-case situations occur, making this approach (as well as exact robust programming) overly conservative.

5 Conclusions

In this paper, we propose an alternative methodology to handle data uncertainty in Kidney Exchange Programs. We developed an interactive tool that visually supports decision makers in the process of choosing the best solution, attending their interests. For that, we calculate the predicted performance of a considerable number of solutions for a set of scenarios generated by Monte Carlo Tree Search where some pairs become unavailable (i.e., failures occur).

A main contribution of this work is to provide DMs with additional information on the decision possibilities, if failures occur. This contrasts with other work in the same area where, in general, only a solution that results in the maximum expected number of transplants is provided.

Acknowledgements This work is financed by the ERDF European Regional Development Fund through the Operational Programme for Competitiveness and Internationalisation - COMPETE 2020 Programme, and by National Funds through the Portuguese funding agency, FCT - Fundação para a Ciência e a Tecnologia, within project "mKEP - Models and optimisation algorithms for multicountry kidney exchange programs" (POCI-01-0145-FEDER-016677), and by COST Action CA15210, ENCKEP, supported by COST (European Cooperation in Science and Technology) – http://www.cost.eu/.

References

1. Abraham, D, Blum, A., Sandholm, T.: Clearing algorithms for Barter exchange markets: enabling nationwide kidney exchanges. In: Proceedings of the 8th ACM conference on Electronic commerce, June 13–16, pp. 295–304 (2007)
2. Bray, M., Wang, W., Song, P.X.-K., Kalbfleisch, J.D.: Valuing sets of potential transplants in a kidney paired donation network. Stat. Biosci. (2018)
3. Bray, M., Wang, W., Song, P.X.K., Leichtman, A.B., Rees, M., Ashby, V.B., Eikstadt, R., Goulding, A., Kalbfleisch, J.: Planning for uncertainty and fallbacks can increase the number of transplants in a kidney-paired donation program. Am. J. Transplant. **15**, 08 (2015)
4. Browne, C., Powley, E., Whitehouse, D., Lucas, S., Cowling, P., Rohlfshagen, P., Tavener, S., Perez Liebana, D., Samothrakis, S., Colton, S.: A survey of Monte Carlo tree search methods. IEEE Trans. Comput. Intell. AI Games **4**(1), 1–43 (2012)
5. Chen, Y., Li, Y., Kalbfleisch, J.D., Zhou, Y., Leichtman, A., Song, P.X.-K.: Graph-based optimization algorithm and software on kidney exchanges. IEEE Trans. Biomed. Eng. **59**(7), 1985–1991 (2012)
6. Constantino, M., Klimentova, X., Viana, A., Rais, A.: New insights on integer-programming models for the kidney exchange problem. Eur. J. Oper. Res. **231**(1), 57–68 (2013)
7. Dickerson, J., Procaccia, A., Sandholm, T.: Failure-aware kidney exchange. In: EC-13: Proceedings of 14th ACM Conference on Electronic Commerce, June, 2013
8. Edmonds, J.: Paths, trees, and flowers. Can. J. Math. **17**, 449–467 (1965)
9. Glorie, K., Carvalho, M., Constantino, M., Viana, A., Klimentova, X.: Robust models for the kidney exchange problem. Technical report DS4DM-2018-007, Data science for real-time decision-making at Polytechnique Montréal (2018)
10. Klimentova, X., Pedroso, J.P., Viana, A.: Maximising expectation of the number of transplants in kidney exchange programmes. Comput. Oper. Res. **73**, 1–11 (2016)
11. Kocsis, L., Szepesvári, C.: Bandit based Monte-Carlo planning. In: Fürnkranz, J., Scheffer, T., Spiliopoulou, M. (eds.) Machine Learning: ECML 2006, pp. 282–293. Berlin, Heidelberg (2006)
12. Li, Y., Song, P.X.-K., Zhou, Y., Leichtman, A.B., Rees, M., Kalbfleisch, J.: Optimal decisions for organ exchanges in a kidney paired donation program. Stat. Biosci. **6**, 05 (2014)
13. Manlove, D., O'Malley, G: Paired and altruistic kidney donation in the UK: algorithms and experimentation. In: Klasing, R. (ed.) Experimental Algorithms. SEA 2012. Lecture Notes in Computer Science, vol. 7276, pp. 271–282 (2012)
14. Pedroso, J.P.: Maximizing expectation on vertex-disjoint cycle packing. In: Murgante, B. et al. (eds.) ICCSA 2014. Lecture Notes in Computer Science, vol. 8580, pp. 32–46 (2014)
15. Pedroso, J.P., Rei, R.: Tree search and simulation. In: Mujica Mota, M.E.A. (ed.) Applied Simulation and Optimization: In Logistics, Industrial and Aeronautical Practice, pp. 109–131. Springer International Publishing (2015)

16. Rapaport, F.: The case for a living emotionally related international kidney donor exchange registry. Transpl. Proc. **18**, 5–9 (1986)
17. Rei, R.: Monte Carlo tree search for combinatorial optimization. PhD thesis, Faculdade de Ciências, Universidade do Porto (2018)
18. Roth, A., Sönmez, T., Ünver, M.: Efficient kidney exchange: coincidence of wants in markets with compatibility-based preferences. Am. Econ. Rev. **97**(3), 828–851 (2007)
19. Sutton, R., Barto, A.: Reinforcement Learning: An Introduction. MIT Press, Cambridge (2018)

Facing Dynamic Demand for Surgeries in a Portuguese Case Study

Mariana Oliveira and Inês Marques

Abstract Operating rooms (ORs) are major cost centers of a hospital. Moreover, ORs performance have large impact in the workload variability of ORs up- and downstream units. The OR management is increasingly challenging due to population ageing, increasing demand and use of expensive technologies. While surgical demand is increasing, resources are highly restricted and thus ORs need to be efficiently managed. Based on a collaboration with a Portuguese public hospital, this paper proposes a dynamic approach for a master surgery scheduling problem to face the increasing surgical demand. Considering space and staff restrictions of the hospital, this work aims to reallocate OR time among the surgical services in order to increase surgical access through matching the available OR capacity with demand. A mathematical programming model is proposed to comply with staff preferences in the slots' allocations, to match surgical supply and demand, and to level the workforce in the up- and downstream units. The model suggests new master surgical schedules, using information and data collected in the hospital under study. Results show that the main bottlenecks are the workforce availability and the stability requirement.

Keywords OR in health services · Operating rooms · Master surgery schedule · Tactical decisions · Dynamic demand · Mathematical model

1 Introduction

Health care organizations face increasingly complex challenges due to the ageing of the worldwide population, the increasing demand for health services and the development of new and expensive methods, equipments and instruments. Operating rooms (ORs) represent almost half of the hospital revenues and costs. Surgical activity is

M. Oliveira (✉) · I. Marques
Centre for Management Studies, Instituto Superior Técnico, University of Lisbon,
Av. Rovisco Pais 1, 1049-001 Lisbon, Portugal
e-mail: mariana.m.oliveira@tecnico.ulisboa.pt

I. Marques
e-mail: ines.marques.p@tecnico.ulisboa.pt

S. Relvas et al. (eds.), *Operational Research*, Springer Proceedings
in Mathematics & Statistics 374, https://doi.org/10.1007/978-3-030-85476-8_7

indeed often considered the engine of the organization [5, 7]. However, OR management can be a very complex process. Besides having a direct and important impact in the health status of patients waiting for a surgery, it also involves a high level of variability and uncertainty. To achieve a higher service level, the patient should wait for a surgery for a lower number of days than a predetermined maximum waiting time meaning that waiting longer is highly undesirable. Thus, the OR time assignment to specialties should follow as much as possible the fluctuations in surgical demand pattern for the various specialties. Moreover, a mismatch of supply and demand might increase the waiting list for some specialties while other specialties might have a low OR utilization of their allocated OR time. Therefore, this paper aims to develop an approach for OR planning to match the changing demand patterns. A mathematical model is proposed to tackle the problem of matching supply and demand, at a tactical level of decision, while considering staff preferences and balancing the utilization of up- and downstream resources.

There are three different main strategies to manage OR time: a flexible approach—open scheduling [4]; a stable approach—block scheduling [5, 6, 17]; and an intermediate—modified block scheduling—which tries to combine both flexibility and stability [1, 8]. Under open scheduling, each time slot (combination of an OR, a day and a time period) is not reserved for any particular surgeon or group of surgeons. Thus, this strategy allows time slots to be used according to the needs, which increases flexibility to cope with changes in demand. Liu et al. state that, especially in large-scale OR cases, open scheduling can increase OR efficiency and decrease overtime costs [11]. For block scheduling, time slots are assigned to a specialty or to a surgeon group [16]. This approach is the one followed by a master surgery schedule (MSS) and provides higher stability to both managers and medical staff, which allows to have a more predictable pattern of bed occupancy in the hospital's units, and also in the required staff and material [4]. This work follows a block scheduling strategy. Under this scope, many objectives and solutions can be applied. On the one hand, Blake and Donald allocate time slots to surgical groups by guaranteeing an equitable distribution [7]. On the other hand, Agnetis et al. adapt the time allocation to each surgical group according to the demand [4]. But, both of these papers consider the OR as an isolated unit in the hospital. Recently, authors started considering downstream units (e.g. [5, 17]). These units have a high impact on the management of the ORs because they can limit the number of operated patients in each day. Moreover, Zhang et al. analyze the impact that ORs can also have in the upstream units, by minimizing the length of stay in the wards before surgery [18]. Mannino et al. solve the MSS problem by proposing a mixed integer linear programming model to balance queues among specialties [13]. Queues are measured by the number of unscheduled patients, which is the difference between the expected demand and the number of scheduled patients. Furthermore, Guido and Conforti discuss trade-offs among underutilization of OR capacity, balanced distribution of OR time among surgical groups, waiting time and overtime [10]. Abedini et al. propose an integer programming model to minimize the number of "blockings" between consecutive units [3]. When there are no available downstream resources, the patient is held ("blocked") in the current stage, which increases waiting time in each stage, overall time in the peri-operative process, overtime and overnight shifts. In 2019, Marques et al. develop a multi-objective mixed integer programming model to build cyclic MSSs, for which time slots can be either assigned to surgical specialties or to individual surgeons [14].

This paper proposes a mathematical model that combines three objectives that have been already studied in the literature, but that have not been studied together. The objectives are: to comply with the surgeons' and anesthesiologists' preferences (e.g. [15]), to allocate slots following demand pattern (e.g. [12]), and to balance workload in up- and downstream units (e.g. [9]). Besides being a general model that can be applied in other contexts, in this paper, the model is validated with and applied to the case of a Portuguese public hospital and real data from the hospital is used to perform the computational experiments. As impacting the agenda of doctors and services, it is common to see in practice almost unchanged MSSs for more than 30 years, regardless the constant changes in waiting lists and surgical demand pattern. Thus, hospitals need to adapt surgical supply to the dynamic demand to be able to manage ORs in the most efficient way and be effective in achieving the goals negotiated with the governing Ministry.

The remainder of this paper is structured as follows. Section 2 details the MSS problem and introduces the notation and the model formulation. In Sect. 3, the model is applied to the case study and results are discussed. Finally, Sect. 4 concludes the paper.

2 Mathematical Model

A mathematical model is proposed to generate new MSSs for a large planning horizon. Schedules are divided in sets of working days (e.g. Monday to Friday) that can be organized in weeks. Each working day is divided in shifts, for which there is a set of available ORs. Thus, to construct the MSSs, each surgical service is assigned to some slot (shift, day and OR). For the sake of simplicity and following the case under study, it is assumed that every slot has the same duration. A weekly target time is stated for each specialty, based on the expected number of patients on the waiting list, the average surgery duration and the total available OR time. Nonetheless, when it is not possible to achieve the target values, the MSS has under or overallocation which is penalized and minimized in the objective function. Furthermore, to establish workload balance among staff members, a maximum and minimum values are forced for the number of slots assigned to each doctor and anesthesiologist and to each service, respectively. Moreover, the availability and preferences of surgeons and anesthesiologists must be considered. However, it is also necessary to guarantee that each assigned slot has the minimum requested number of available surgeons and anesthesiologists to work. Additionally to ORs, up- and downstream units and their relative importance are taken into account. Resources as the pre-ward, the intensive-care unit (ICU) and wards have restricted capacity, which is considered when allocating OR slots to the services. For each unit, a target value for the number of occupied beds is also set to balance workload in these units and avoid variability, and thus under and overutilization of up- and downstream units should be minimized. Indeed, the number of occupied beds depends not only in the allocated slots but also on estimated probabilities for patients of each specialty to be in the unit in a certain day after being operated. In order to increase acceptability of these approaches, staff satisfaction and MSS stability should be considered. Accordingly, to match demand variations, a non-cyclic MSS is allowed. However, to guarantee staff routines a maximum number of monthly (relatively to its corresponding week of the first month) and weekly (relatively to the first week of the considered month) changes is stated. These numbers should balance both MSS stability and flexibility to cope with changes in the demand pattern.

To solve this MSS problem, a mathematical model is proposed in (1)–(25). The indices, sets, subsets, parameters, decision and auxiliary variables are presented in Table 1. Constraints (2)–(10) formulate the functionality of the ORs, namely the staff availability and workload. Moreover, constraints (11)–(14) model the demand and supply for allocated slots. Constraints (15)–(18) guarantee the MSS monthly and weekly stability. Finally, constraints (19)–(23) deal with the workload in the up- and downstream units.

Function (1) comprises three objectives: the first one maximizes the number of slots assigned to each specialty, weighted by the relative aggregate preferences of staff members (surgeons and anesthesiologists); the second and third objectives minimize undesired deviations to target values for the OR time assigned to each specialty and the number of occupied beds in the up- and downstream units. The workload in these units is weighted by a parameter which represents a relative importance of each unit and by the target value for beds' occupation.

$$\max \quad \sum_{s \in S} \sum_{w \in W} \sum_{d \in D} \sum_{b \in B} \sum_{r \in R} \left(\frac{\sum_{i \in I_s} \kappa_{idb}^{surg}}{|I|} + \frac{\sum_{a \in A} \kappa_{adb}^{anest}}{|A|} \right) x_{swdbr} - \frac{1}{|W|} \sum_{s \in S} \sum_{w \in W} (t_{sw}^{-} + t_{sw}^{+}) - $$
$$- \sum_{z \in Z} w_z \sum_{k \in K} \frac{u_{zk}^{-} + u_{zk}^{+}}{u_{zk}} \tag{1}$$

$$\text{s.t.:} \quad \sum_{s \in S} x_{swdbr} \leq 1 \quad \forall w \in W, d \in D, b \in B, r \in R \tag{2}$$

$$\sum_{s \in S} \sum_{d \in D} \sum_{b \in B} \sum_{r \in R} x_{swdbr} \leq slots_w \quad \forall w \in W \tag{3}$$

$$\delta^{surg} \sum_{r \in R} x_{swdbr} \leq a_{swdb}^{surg} \quad \forall s \in S, w \in W, d \in D, b \in B \tag{4}$$

$$\delta^{surg} \sum_{b \in B} \sum_{r \in R} x_{swdbr} \leq \sum_{i \in I_s} a_{iwd}^{surgD} \quad \forall s \in S, w \in W, d \in D \tag{5}$$

$$\delta^{surg} \sum_{d \in D} \sum_{b \in B} \sum_{r \in R} x_{swdbr} \leq \sum_{i \in I_s} ww_i^{surg} \quad \forall s \in S, w \in W \tag{6}$$

$$\delta^{anest} \sum_{s \in S} \sum_{r \in R} x_{swdbr} \leq a_{wdb}^{anest} \quad \forall w \in W, d \in D, b \in B \tag{7}$$

$$\delta^{anest} \sum_{s \in S} \sum_{b \in B} \sum_{r \in R} x_{swdbr} \leq \sum_{a \in A} a_{awd}^{anestD} \quad \forall w \in W, d \in D \tag{8}$$

Table 1 Indices, sets, subsets, parameters and variables for the mathematical model

Indices and sets	
$s \in S$	specialties
$m \in M$	months
$w \in W$	weeks
$d \in D$	weekly working days
$k \in K$	days in the planning horizon; the first day of the planning horizon is $k = 1$
$r \in R$	operating rooms
$b \in B$	shifts
$i \in I$	surgeons
$a \in A$	anesthesiologists
$z \in Z$	up- and downstream units
Subsets	
W_m	weeks of month m; the first week of month m is w_{1m}
S_z	specialties that use unit z
I_s	surgeons of specialty s
Parameters	
$slots_w$	total number of time slots (combination of day d, shift b and OR r) in week w
θ	duration of each slot (in hours)
κ_{idb}^{surg}	preference score for surgeon i to work on shift b on day d
κ_{adb}^{anest}	preference score for anesthesiologist a to work on shift b on day d
w_z	relative weight of unit z
a_{swdb}^{surg}	number of surgeons of specialty s available on week w, day d and shift b
a_{iwd}^{surgD}	1, if surgeon i is available on at least one shift on week w and day d; 0, otherwise
a_{wdb}^{anest}	number of anesthesiologists available on week w, day d and shift b
a_{awd}^{anestD}	1, if anesthesiologist a is available on at least one shift on week w and day d; 0, otherwise
$inic_s$	number of patients of specialty s in the waiting list in the first day of the planning horizon
ent_{sw}	number of patients of specialty s entering the waiting list on week w
ww_i^{surg}	maximum weekly workload for surgeon i
ww_a^{anest}	maximum weekly workload for anesthesiologist a
mw_{sm}	minimum workload for specialty s on month m
Δ_m^M	monthly stability for month m
Δ_w^W	weekly stability for week w
e_{zsk}	probability that a patient of specialty s is in unit z on day k before (for upstream units) or after (for downstream units) the surgery
dur_s	average duration of a surgery of specialty s (in hours)
λ_s	average number of patients operated per slot by specialty s
n_{zs}	maximum number of days that a patient of specialty s stays in unit z

(continued)

Table 1 (continued)

Indices and sets	
u_{zk}	target utilization for unit z on day k
c_{zk}	available capacity of unit z on day k
δ^{surg}	minimum requested number of surgeons available per slot assigned
δ^{anest}	minimum requested number of anesthesiologists available per slot assigned
G	large number
Decision variables	
x_{swdbr}	1, if specialty s is assigned to OR r on week w, day d and shift b; 0, otherwise
t^-_{sw}, t^+_{sw}	negative and positive deviations of the allocated time to the target value for specialty s on week w, respectively
u^-_{zk}, u^+_{zk}	under and overutilization of beds on unit z on day k (compared to the target utilization value), respectively
Auxiliary variables	
t_{sw}	target time allocation for specialty s in week w
y_{swdbr}	0, if specialty s is assigned on week w to the same OR r, day d and shift b as the first week of the same month; 1, otherwise
j_{swdbr}	0, if specialty s is assigned on week w to the same OR r, day d and shift b as in the corresponding week on the first month of the planning horizon; 1, otherwise
f_{zk}	expected number of patients in unit z on day k
v^{t}_{sw}	0, if $t^-_{sw} > 0$; 1, if $t^+_{sw} > 0$
v^{u}_{zk}	0, if $u^-_{zk} > 0$; 1, if $u^+_{zk} > 0$
p_{sw}	number of patients of specialty s on the waiting list in the beggining of week w

$$\delta^{anest} \sum_{s\in S} \sum_{d\in D} \sum_{b\in B} \sum_{r\in R} x_{swdbr} \le \sum_{a\in A} ww_a^{anest} \quad \forall w \in W \tag{9}$$

$$\sum_{w\in W_m} \sum_{d\in D} \sum_{b\in B} \sum_{r\in R} x_{swdbr} \ge mw_{sm} \quad \forall s \in S, m \in M \tag{10}$$

$$p_{sw} = p_{s,w-1} + ent_{s,w-1} - \sum_{d\in D} \sum_{b\in B} \sum_{r\in R} \lambda_s x_{s,w-1,d,b,r} \quad \forall s \in S, w \in W \setminus \{1\} \tag{11}$$

$$p_{s1} = inic_s \quad \forall s \in S \tag{12}$$

$$t_{sw} = p_{sw} dur_s \quad \forall s \in S, w \in W \tag{13}$$

$$\theta \sum_{d\in D} \sum_{b\in B} \sum_{r\in R} x_{swdbr} + t^-_{sw} - t^+_{sw} = t_{sw} \quad \forall s \in S, w \in W \tag{14}$$

$$|x_{swdbr} - x_{sw_{1m}dbr}| = y_{swdbr} \quad \forall s \in S, w \in W_m \setminus \{w_{1m}\}, m \in M, d \in D, b \in B, r \in R \tag{15}$$

$$\sum_{s\in S} \sum_{d\in D} \sum_{b\in B} \sum_{r\in R} y_{swdbr} \le \Delta_w \quad \forall w \in W \tag{16}$$

$$|x_{swdbr} - x_{s1dbr}| = j_{swdbr} \quad \forall s \in S, w \in W_m, m \in M \setminus \{1\},$$

$$l = w - \sum_{g<m} |W_g|, d \in D, b \in B, r \in R \tag{17}$$

$$\sum_{s\in S} \sum_{w\in W_m} \sum_{d\in D} \sum_{b\in B} \sum_{r\in R} j_{swdbr} \leq \Delta_m \quad \forall m \in M \tag{18}$$

$$0 \leq f_{zk} - \sum_{s\in S_z} \sum_{b\in B} \sum_{r\in R} \sum_{l=0}^{n_{zs}-1} \lambda_s e_{zsk} x_{s,w,d\pm l,b,r} \leq 1 \quad \forall z \in Z, k \in K : k \to (w, d),$$
$$w \in W, d \in D \tag{19}$$

$$f_{zk} + u_{zk}^{-} - u_{zk}^{+} = u_{zk} \quad \forall z \in Z, k \in K \tag{20}$$

$$u_{zk}^{+} \leq c_{zk} - u_{zk} \quad \forall z \in Z, k \in K \tag{21}$$

$$u_{zk}^{-} \leq G\left(1 - v_{zk}^{u}\right) \quad \forall z \in Z, k \in K \tag{22}$$

$$u_{zk}^{+} \leq G\, v_{zk}^{u} \quad \forall z \in Z, k \in K \tag{23}$$

$$t_{sw}^{-}, t_{sw}^{+}, u_{zk}^{-}, u_{zk}^{+}, f_{zk} \geq 0 \quad \forall s \in S, w \in W, z \in Z, k \in K \tag{24}$$

$$x_{swdbr}, y_{swdbr}, j_{swdbr}, v_{sw}^{t}, v_{zk}^{u} \in \{0, 1\} \quad \forall s \in S, w \in W, d \in D, b \in B, r \in R,$$
$$z \in Z, k \in K \tag{25}$$

Constraints (2) state that a maximum of one specialty must be assigned to each slot available (block scheduling approach). Besides having only a specialty per slot, schedules must be constructed considering the maximum number of available slots in each week. Thus, constraints (3) state the overall weekly number of assigned slots. Constraints (4) guarantee that a slot is allocated to a specialty only if there is a minimum number of available surgeons. These are important constraints, because in many cases, a surgery must be performed by more than one surgeon. For example, the Portuguese legislation states that two surgeons must be in the OR for a surgery to be performed [2]. Constraints (5) avoid having a surgeon allocated in consecutive slots, by guaranteeing that, in a day, the number of slots assigned to some specialty is not higher than the feasible number of slots considering surgeons' availability and the minimum number of doctors per surgery. Moreover, constraints (6) restrict the weekly workload for surgeons. Similar constraints are defined to anesthesiologists in (7)–(9). Moreover, a minimum monthly workload for each specialty is also assigned with (10). Constraints (11) estimate the number of patients of each specialty, waiting for a surgery, at the beginning of each week based on the number of patients that were on the waiting list at the beginning of the previous week, the number of estimated new entries during the week, and the average number of performed surgeries on the previous week. Constraints (12) state the specific case of the first week of the planning horizon. Based on the dynamic number of patients in the waiting list for each specialty and week, and considering the average duration of each type of surgeries, constraints (13) assign a target value for the allocated OR time to each specialty and week measured by the estimated time required to perform all the surgeries waiting. Constraints (14) define the positive and negative deviations to the OR time target value to each specialty on a certain week as a result of the allocation of slots. Note that in the case of OR time as a scarce resource (as is common among the Portuguese cases), the deviations will tend to be negative. Auxiliary variables to measure the (weekly and monthly) stability of the MSSs, y_{swdbr} and j_{swdbr}, are defined respectively by constraints (15) and (17). To simplify the reading of the model, the original non-linear constraints are presented although these constraints were linearized for the computational experiments. Moreover, to guarantee staff routines through MSS stability, constraints (16) and (18) establish the maximum number of weekly and monthly changes, respectively. Constraints (19) define the daily number of occupied beds in each unit based on variables x. Three notes about these constraints. First, since probabilities are continuous numbers, these constraints cannot be linearly written in equality. Second, (w, d) time measure needs to be transformed in continuous days (k) since up- and downstream resources do not have non-working days (unlike the OR). Finally, $d - l$ and $d + l$ are used for up- and downstream units, respectively. Constraints (20) define positive and negative deviations of the expected number of occupied beds to the target beds utilization. Constraints (21) limit the positive deviation for the expected number of occupied beds in each unit. Moreover, constraints (22) and (23) force that if the beds utilization positive (or negative) deviation is not zero, than the negative (or positive) deviation should be zero. Although the minimization of the third term of the objective function (if a single model objective) does not require these constraints, preliminary results for model validation showed that the multi-objective effect requires these additional constraints. Finally, constraints (24) and (25) establish the domain sets for decision and auxiliary variables.

Although this mathematical model is motivated by a Portuguese public hospital, it aims to be as general as possible. The proposed approach can be used to tackle the very common problem of not considering demand and workload of other facilities in OR planning. The resulting MSSs should maximize the number of allocated slots to each specialty, while considering surgeons and anesthesiologists' preferences, target values and up- and downstream units' workload.

3 Computational Experiments

The proposed mathematical model is applied in the context of a public hospital in Portugal. For this specific central surgical suite, the MSS is stable among the past years. However, surgeries demand and availability of staff change over time and therefore dynamic models must be created. The patients follow a certain path from admission to discharge and may use six up- and downstream units ($|Z| = 6$): a pre-ward, the ICU and four wards. Before surgery, each patient stays in a pre-ward for a maximum of one day. After leaving the OR, the patient can go to the ICU or to one of the 4 wards, depending on his/her condition. In this hospital, there are 4 ORs ($|R| = 4$) to be used for elective surgeries. These ORs are shared by 8 surgical specialties ($|S| = 8$), namely general surgery, urology, orthopedics, ophthalmology, plastic surgery, pediatric surgery, otorhinolaryngology (ORL) and stomatology. Elective surgeries are performed from Monday to Friday ($|D| = 5$). Each working day is divided in two shifts ($|B|=2$), a morning and an afternoon shift. The surgeries are performed by

Table 2 Results for the case study. Time limit: 30 min

Instance	Obj1Value	Obj2Value	Obj3Value	Gap	Time
RC	1357	7532.05	550.92	0.84	30
IC	2130	5416.67	481.39	2.34	30

forty-three surgeons ($|I| = 43$) and ten anesthesiologists ($|A| = 10$). Furthermore, surgeons divide their working time among ORs, appointments, internment visits and emergency cases and, therefore, a maximum of one or two weekly slots must be allocated to each surgeon ($ww_i^{surg} = 1, 2$). In addition to OR related tasks, anesthesiologists perform appointments and thus the maximum number of weekly allocated slots is four or five ($ww_a^{anest} = 4, 5$). Surgeons and anesthesiologists time available for OR is therefore limited. Moreover, each specialty must receive at least one slot per month ($mw_{sm} = 1$). For a surgery to occur, a minimum number of surgeons and anesthesiologists must be in the OR. According to Portuguese legislation [2], a minimum of two surgeons ($\delta^{surg} = 2$) and one anesthesiologist ($\delta^{anest} = 1$) must participate in each surgery. A one-year planning horizon ($|M| = 12$; $|W| = 52$; $|K| = 365$) is considered.

The model is implemented in Java, with Eclipse IDE 2018-12, using the callable library of ILOG CPLEX 12.8.0. All tests ran on a portable computer with an Intel Core i7-8550U CPU of 1.80 GHz and 8 GB of RAM under Windows 10 operating system. A time limit of 30 min was used for the computational experiments.

Based on the real data from the hospital under study, two instances were used for the computational experiments. Instance 'RC' uses the full set of real parameters; and instance 'IC' corresponds to the same situation as 'RC' but the work capacity of each surgeon in the OR is increased by 2 weekly slots. The results obtained are summarized in Table 2. 'Obj1Value', 'Obj2Value' and 'Obj3Value' show, respectively, the value for the first, second and third terms of the objective function. Thus, 'Obj1Value' is the number of slots assigned weighted by the relative aggregate preferences of surgeons and anesthesiologists. 'Obj2Value' measures the deviations of the number of slots assigned to each specialty with respect to the target value set by the dynamic waiting list for surgery. 'Obj3Value' are the deviations to the target value for the number of occupied beds in the up- and downstream units. 'Gap' is an upper bound on the optimality gap (in %) based on the LP relaxation solved by CPLEX. Finally, 'Time' is the computation time (in minutes). Table 2 shows that increasing workload capacity of surgeons for ORs can have a high impact in the value of the each individual term of the objective function. Indeed, besides attaining a higher value for the compliance of aggregated preferences ('Obj1Value') and making it easier to achieve the beds' utilization targets ('Obj3Value') for the same allocation target values, each surgeon can perform more surgeries and therefore minimize the allocation deviations ('Obj2Value'). Thus, results show that the best compliance with surgeries dynamic demand is obtained with 'IC' instance.

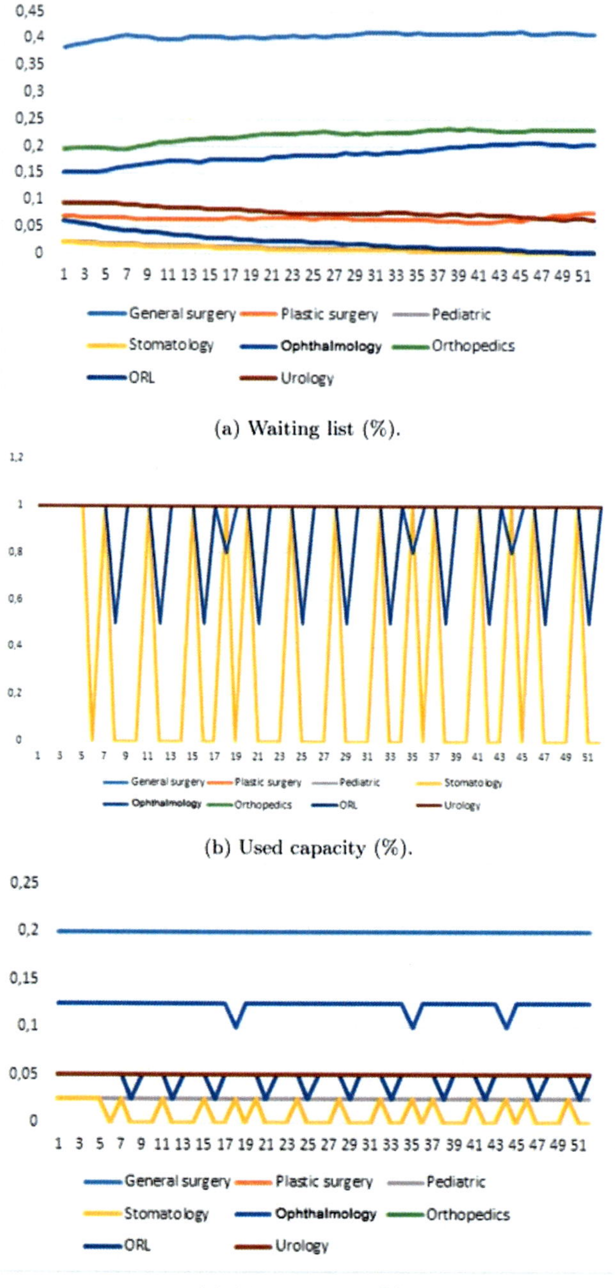

Fig. 1 Results for instance 'RC'

Fig. 2 Results for instance 'IC'

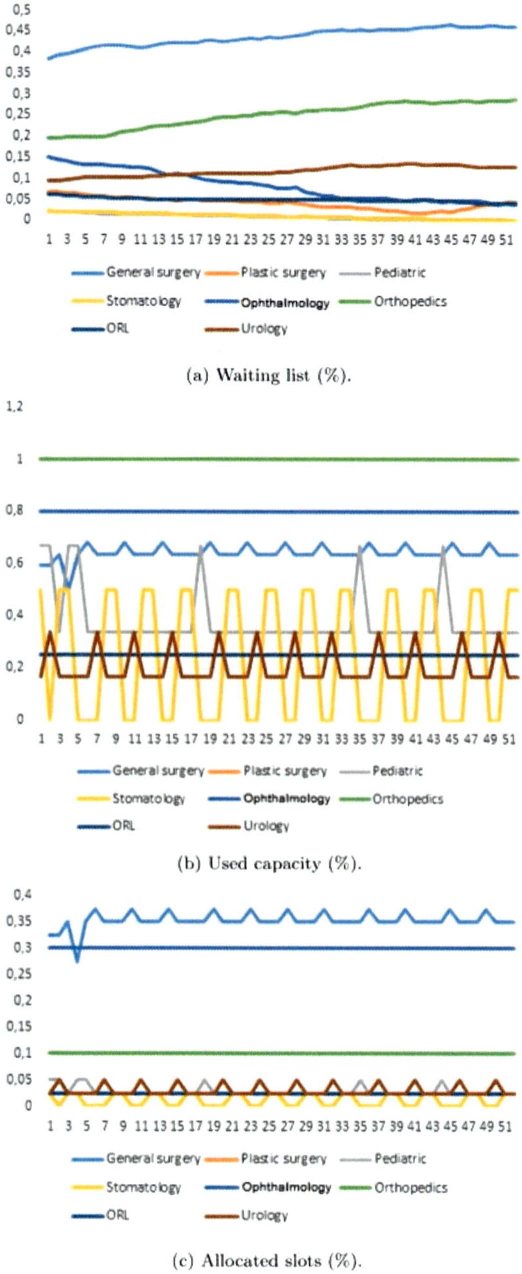

(a) Waiting list (%).

(b) Used capacity (%).

(c) Allocated slots (%).

Figures 1 and 2 show results for 'RC' and 'IC' instances, respectively: (a) the percentage of demand of each specialty with respect to the total surgical demand (here, demand is measured in terms of expected duration of surgeries instead of in number of surgeries); (b) the percentage of weekly allocated slots for each specialty when compared to the maximum number of slots that can be allocated to the specialty considering workforce, staff availability and the remaining model constraints ("Used capacity"); and (c) the percentage of allocated slots compared to the total number of available slots for every specialty. Figure 1 shows that, although most specialties use their maximum capacity, there is a low compliance with the demand and an average of 55.24% of the weekly total slots capacity is used. Ophthalmology, ORL and stomatology are the only specialties that do not use their maximum capacity in some weeks. However, these latter two specialties, together with pediatric surgery, are the ones reaching a lower percentage of demand at the end of the planning horizon. Moreover, ophthalmology is one of the specialties that requires more beds in the up- and downstream units per each assigned slot. This, together with the doctors' preferences, leads to the fluctuation in the used capacity. Regarding 'IC' instance, it is possible to increase the used capacity to 95.9% by increasing the surgeons' workload capacity. These results allow to conclude that current workforce capacity is a bottleneck to the effectiveness of the OR and changing the capacity introduces more flexibility to all specialties. Moreover, the compliance with the dynamic demand is not limited by the low capacity of the service. Thus, increasing surgeons work capacity in the ORs is strongly suggested. Figure 2 shows that the main difference between demand and supply is obtained for orthopedics. This is a critical specialty because there are only three available surgeons to perform surgeries in this service and the surgeries are usually longer compared to other specialties. However, this is one of the specialties with a higher number of patients in the waiting list at the beginning of the planning horizon. Thus, although using full capacity, it is very difficult to significantly reduce the waiting list.

Additionally, stability constraints can highly restrict an effective response to fluctuations in demand pattern. To analyze this effect, tests for 'IC' instance were repeated removing constraints (15)–(18) which model the MSS monthly and weekly stability request, respectively. Results are shown in Fig. 3. Indeed, by comparing the three graphs one can conclude that the pattern of slots' allocation can better follow the demand patterns. For this reason, it is safe to state that, without the stability requirement, the surgical supply is much easier adjusted to follow the demand dynamic.

Moreover, Table 3 analyses the impact of MSS stability on the number of patients waiting for surgery, for each specialty, at the end of the planning horizon, in both 'RC' and 'IC' instances: $inic_s$ is the number of patients in the waiting list at the beginning of the planning horizon; and 'wS' and 'woS' show the number of patients in the waiting list at the end of the planning horizon, with or without considering the stability constraints, respectively. Comparing the results obtained for the real case ('RC' instance), some differences can be spotted. Stomatology, ophthalmology and ORL can schedule more patients, in one year, when stability is not considered (73 additional patients in total): 25 more patients for stomatology (waiting list length decreases from 27 to 2 patients); 18 more patients for ophthalmology (decrease from

Fig. 3 Results for instance 'IC', without stability constraints

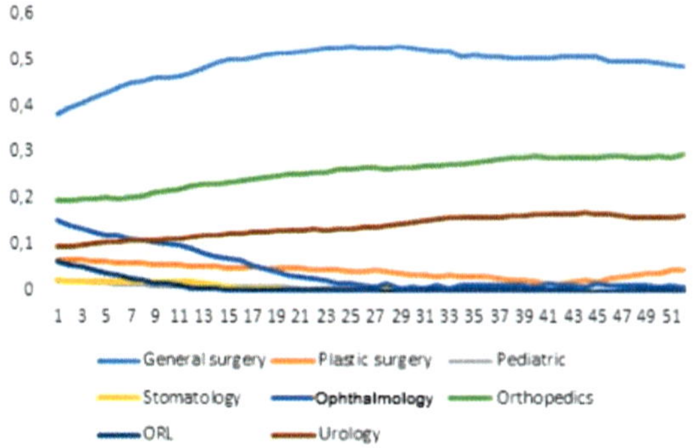

(a) Waiting list (%).

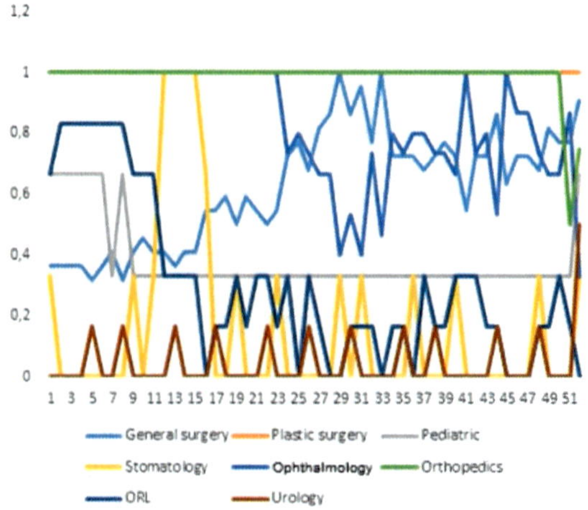

(b) Used capacity (%).

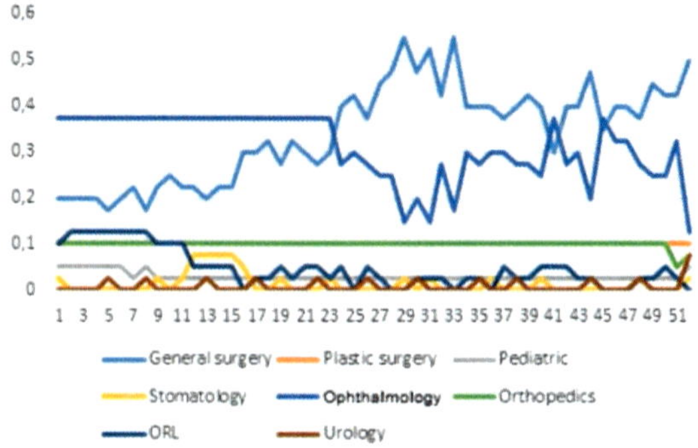

(c) Allocated slots (%).

Table 3 Results with ('wS') and without ('woS') stability constraints

Specialty	$inic_s$	'RC'		'IC'	
		wS	woS	wS	woS
General	888	2073	2073	1449	1497
Plastic	254	623	623	215	215
Pediatric	161	70	70	7	7
Stomatology	100	27	2	2	2
Ophthalmology	825	2431	2413	271	61
Orthopedics	514	1326	1326	1020	1026
ORL	260	30	0	230	5
Urology	230	343	343	421	525
Total	3232	6923	6850	3615	3338

2431 to 2413 patients); and 30 more surgeries for ORL (waiting list decreases from 30 to 0). However, the impact of not considering stability is higher when more workload capacity is available ('IC' instance, 277 patients in total). On the one hand, for general surgery, orthopedics and urology slightly worst results are obtained when stability constraints are not considered, meaning that longer waiting lists are obtained in the end of the planning horizon: general surgery waiting list increases by 48 patients (from 1449 to 1497); orthopedics waiting list has 6 more patients (from 1020 to 1026); and urology waiting list has 104 more patients (from 421 to 525 patients). On the other hand, the additional surgeries performed for ophthalmology and for ORL outperforms the latter specialties increasing the overall impact in this case: ophtalmology service can perform 210 more surgeries (waiting list decreases from 271 to 61) and ORL performs 225 more surgeries (waiting list decreases from 230 to 5 patients). In general, having no stability constraints gives more flexibility to adapt to the dynamic demand, allowing the hospital to perform more surgeries, both for 'RC' and 'IC' instances. Thus, both 'RC' and 'IC' instances obtain better results regarding the value of the second term of the objective function (7467.95 and 5256.41, respectively) when the stability constraints are not considered.

4 Conclusions

This paper proposes a mathematical model for a master surgery scheduling problem motivated by the case of a Portuguese public hospital. The problem considers a large planning horizon that enables to adjust to fluctuations in surgical demand pattern. This is performed by defining a target OR time to each specialty in each week which depends on (1) the number of new entries in each week (new demand), (2) the balance of the number of patients in the waiting list resulting from decisions in previous weeks (demand not met in previous periods), and (3) on the estimated time required for the patients waiting for surgery. The model integrates three objectives that are not usually considered together: maximize the number of OR slots assigned weighted by workforce preferences; adapt the supply to dynamic fluctuations in demand patterns; and balance workload in up- and downstream units. The model generalizes tactical decisions in OR management and the case of the Portuguese public hospital is analyzed in the computational experiments using real data from the hospital. Results show that the main drawbacks in complying with the dynamic demand fluctuations are the workforce and the stability constraints imposed by the model and valued by the OR staff and main stakeholders.

For future work, a sensitivity analysis to the stability parameters is suggested. This parameter sets the trade-off between stability and flexibility. Furthermore, a more detailed discussion on how to consistently measure the stakeholders' preference should be considered. Predictive models should be developed to forecast weekly demand and simulation models should also be used to investigate the impact of the generated MSSs on the patients' scheduling and on the OR efficiency and effectiveness, such as case cancellation rates or prolonged turnover times.

Acknowledgements The authors acknowledge the support provided by FCT and P2020 under the project PTDC/EGE OGE/30442/2017, Lisboa-01.0145-Feder-30442.

References

1. Abdelrasol, Z., Harraz, N., Eltawil, A.: Operating room scheduling problems: a survey and a proposed solution framework. In: Kim, H.K., Ao, S.-I., Amouzegar, M.A. (eds.) Transactions on Engineering Technologies, pp. 717–731. Springer, Dordrecht (2014)
2. ERS: Deliberação do Conselho de Administração da Entidade Reguladora da Saúde (ERS). Entidade Reguladora da Saúde (2016). https://www.ers.pt/uploads/document/file/14103/Vers_o_n_o_Confidencial__ERS_61-17.pdf
3. Abedini, A., Li, W., Ye, H.: An optimization model for operating room scheduling to reduce blocking across the perioperative process. Procedia Manuf. **10**, 60–70 (2017)
4. Agnetis, A., Coppi, A., Corsini, M., Dellino, G., Meloni, C., Pranzo, M.: Long term evaluation of operating theater planning policies. Oper. Res. Health Care **1**(4), 95–104 (2012)
5. Beliën, J., Demeulemeester, E.: Building cyclic master surgery schedules with leveled resulting bed occupancy. Eur. J. Oper. Res. **76**(2), 1185–1204 (2007)
6. Blake, J.T., Dexter, F., Donald, J.: Operating room managers' use of integer programming for assigning block time to surgical groups: a case study. Anesth. Analg. **94**(1), 143–148 (2002)
7. Blake, J.T., Donald, J.: Mount Sinai hospital uses integer programming to allocate operating room time. Interfaces **32**(2), 63–73 (2002)
8. Day, R., Garfinkel, R., Thompson, S.: Integrated block sharing: a win-win strategy for hospitals and surgeons. Manuf. Serv. Oper. Manag. **14**(4), 567–583 (2012)
9. Dellaert, N., Jeunet, J.: A variable neighborhood search algorithm for the surgery tactical planning problem. Comput. Oper. Res. **84**, 216–225 (2017)
10. Guido, R., Conforti, D.: A hybrid genetic approach for solving an integrated multi-objective operating room planning and scheduling problem. Comput. Oper. Res. **87**, 270–282 (2017)
11. Liu, Y., Chu, C., Wang, K.: A new heuristic algorithm for the operating room scheduling problem. Comput. Ind. Eng. **61**(3), 865–871 (2011)

12. Malik, M.M., Khan, M., Abdallah, S.: Aggregate capacity planning for elective surgeries: a bi-objective optimization approach to balance patients waiting with healthcare costs. Oper. Res. Health Care **7**, 3–13 (2015)
13. Mannino, C., Nilssen, E.J., Nordlander, T.E.: A pattern based, robust approach to cyclic master surgery scheduling. J. Sched. **15**, 553–563 (2012)
14. Marques, I., Captivo, M.E., Barros, N.: Optimizing the master surgery schedule in a private hospital. Oper. Res. Health Care **20**, 11–24 (2019)
15. Penn, M.L., Potts, C.N., Harper, P.R.: Multiple criteria mixed-integer programming for incorporating multiple factors into the development of master operating theatre timetables. Eur. J. Oper. Res. **262**(1), 194–206 (2017)
16. Samudra, M., Van Riet, C., Demeulemeester, E., Cardoen, B., Vansteenkiste, N., Rademakers, F.E.: Scheduling operating rooms: achievements, challenges and pitfalls. J. Scheduling **19**(5), 493–525 (2016)
17. Santibáñez, P., Begen, M., Atkins, D.: Surgical block scheduling in a system of hospitals: an application to resource and wait list management in a British Columbia health authority. Health Care Manag. Sci. **10**(3), 269–282 (2007)
18. Zhang, B., Murali, P., Dessouky, M.M., Belson, D.: A mixed integer programming approach for allocating operating room capacity. J. Oper. Res. Soc. **60**(5), 663–673 (2009)

Adaptive Sequence-Based Heuristic for Two-Dimensional Non-guillotine Packing Problems

Óscar Oliveira, Dorabela Gamboa, and Elsa Silva

Abstract We present heuristics for two related two-dimensional non-guillotine packing problems. The first problem aims to pack a set of items into the minimum number of larger identical bins, while the second aims to pack the items that generates most value into one bin. Our approach successively creates sequences of items that defines a packing order considering knowledge obtained from sequences previously generated. Computational experiments demonstrated that the proposed heuristics are very effective in terms of solution quality with small computing times.

Keywords Two-dimensional · Rectangular · Non-guillotine · Knapsack problem · Bin packing · Heuristics

1 Introduction

We present heuristics to solve two related non-guillotine packing problems both considering a finite set of m rectangular items types with associated length (l_i), height (h_i), value (v_i), and demand (d_i). In the first problem, the set of items must be packed into the minimum number of identical bins, while the second, aims to maximize the value of the items packed into one bin. The items must be packed orthogonally and cannot be rotated. All bins are rectangular and have identical length (L) and height (H). Giving that, the proposed heuristics deal with each demanded item individually, the first problem, following the typology of Wäscher et al. [45], is classified as Single Bin Size Bin Packing problem (SBSBPP) while the second one is classified as Single Knapsack Problem (SKP).

Ó. Oliveira (✉) · D. Gamboa
CIICESI, Escola Superior de Tecnologia e Gestão, Politécnico do Porto, Felgueiras, Portugal
e-mail: oao@estg.ipp.pt

D. Gamboa
e-mail: dgamboa@estg.ipp.pt

E. Silva
INESC TEC, Faculdade de Engenharia, Universidade do Porto, Porto, Portugal
e-mail: emsilva@inesctec.pt

S. Relvas et al. (eds.), *Operational Research*, Springer Proceedings in Mathematics & Statistics 374, https://doi.org/10.1007/978-3-030-85476-8_8

The heuristics proposed in this paper, as most heuristics proposed in the literature for solving these problems, are sequence-based, i.e., heuristics that generate, at each iteration, a new sequence of items that defines the ordering to pack the items. Although simple, the approach taken by the proposed heuristics for solving the above mentioned problems allows good results with small computational times. This paper extends the work presented in [40] where only the first problem is considered.

The remaining of the paper is organized as follows. Section 2 gives an overview of the solutions approaches proposed in the literature to solve both problems. In Sect. 3, a description of our heuristics, hereafter denoted as Adaptive Sequence-based Heuristics (ASH), are presented, and in Sect. 4, computational results are reported comparing our approach to other solution methods from the literature. Finally, conclusions and future work directions are given in Sect. 5.

2 Literature Review

Since the work of Gilmore and Gomory [22] the interest on cutting and packing problems has been growing due to its complexity (NP-hard, see Garey and Johnson [21]) and great practical applicability (see Singh and Jain [43]).

The next two sub-sections present the most relevant solution methods proposed for the SBSBPP and SKP, respectively.

2.1 *Single Bin Size Bin Packing Problem*

Berkey and Wang [5] studied the performance of heuristics for the SBSBPP adapted from heuristics proposed in the literature for the Open Dimensional Problem (ODP, see Wäscher et al. [45]).

Martello and Vigo [37] showed that the Continuous Lower Bound ($CLB = \lceil \frac{\sum_{i=1}^{m} l_i h_i}{LH} \rceil$) for this problem has a worst-case performance ratio of $\frac{1}{4}$ and presented new lower bounds that are used in a Branch-and-Bound algorithm.

Lodi et al. [36] presented heuristics to solve the 2D (non-)oriented (non-) guillotine SBSBPP and an Unified Tabu Search Framework that is adaptable for each specific problem changing uniquely the inner heuristic to explore the neighbourhood.

Boschetti and Mingozzi in [8] presented new lower bounds and in [9] presented a heuristic, called HBP, for the SBSBPP. HBP generates solutions considering the current allocation method and updating the value of the items at the end of each iteration.

Faroe et al. [14] presented a heuristic algorithm based on Guided Local Search (GLS) for the 3D SBSBPP and evaluate this approach also for the 2D case providing the same depth for bins and items. The algorithm starts with an upper bound on the number of available bins and iteratively decreases this number for the next cutting plan generation. This process is repeated until the time limit is reached or the current solution is the same as the calculated lower bound.

Monaci and Toth [39] presented the Set-Covering Heuristic (SCH) formulating the problem as a Set-Covering Problem. SCH generates, through heuristic procedures, a large set of columns to define the Set-Covering instance, then solved by means of a Lagrangean-based heuristic.

Parreño et al. [41] presented a Greedy Randomized Adaptive Search Procedure (GRASP) with Variable Neighbourhood Descent (VND) for the 2D and 3D SBSBPP.

Blum and Schmid [6] presented a hybrid algorithm, called EA-LGFi, in which an evolutionary approach is used to generate on the input sequences to the LGFi heuristic proposed by Wong et al. [47] in order to perform the placement.

2.2 Single Knapsack Problem

Exact approaches to solve the SKP have been proposed by several authors. Beasley [3] presented a Branch-and-Bound algorithm that uses an upper bound obtained through the Lagrangean relaxation of a 0-1 integer linear programming formulation of the problem and Subgradient Optimization to minimize its value.

Scheithauer and Terno [42] presented an integer programming formulation where binary variables specify the item position relatively to another item. The authors presented a Branch-and-Bound in which Subgradient Optimization is used to minimize the upper bound obtained through the Lagrangean Relaxation of the problem. The authors perform reduction tests that limit the size of the tree, therefore, reducing the computational efforts required by the algorithm.

Fekete and Schepers [15–17] presented a tree-search algorithm for the d-dimensional Knapsack Problem using a graph-theoretical characterization of packings. In this characterization, if no overlapping occurs in both x and y axis projection graphs, the pattern is feasible.

Boschetti et al. [10] presented new upper bounds obtained through the relaxation of their integer programming formulation of the problem.

Caprara and Monaci [11] presented four exact algorithms based on the relaxation of this problem given by the one-dimensional case. Exact algorithms were also proposed for the Knapsack problem in which only guillotine cuts are allowed, without a limitation on the number of stages, in Dolatabadi et al. [13] and more recently in Fleszar [19] and in Furini et al. [20].

Heuristics methods have, also, been proposed to solve the SKP. Lai and Chan [31] presented a heuristic based on the Simulated Annealing using an ordered list of items that encodes the order in which the items will be packed into the bin through a placement algorithm. The placement algorithm packs each item, in turn, without overlapping, into the empty rectangular space (ERS) that is closest to the bin bottom-left corner. This heuristic keeps track of the ERS through a called Difference Process that creates interval lists after each item packing.

Leung et al. [35] performed a comparison of the Genetic Algorithm with bottom-left approach proposed for the Strip Packing Problem by Jakobs [29] and the Simulated Annealing with Difference Process combining their approaches for this problem.

Leung et al. [34] compared a pure Genetic Algorithm and a hybrid approach (Genetic Algorithm with Simulated Annealing) called MSAGA to verify if the latter could prevent the early convergence observed in the pure approach. The decoder, i.e., placement algorithm, uses the Difference Process.

Beasley [4] proposed a Genetic Algorithm based on a new non-linear mathematical formulation.

Alvarez-Valdes et al. presented in [1] a GRASP, and in [2] a Tabu Search to solve this problem.

Hadjiconstantino and Iori [26] presented a greedy heuristic and a hybrid Genetic Algorithm. The greedy heuristic called HCHV packs the items at the bottom-left position possible, alternatively on top of each other or side by side until no more items fit in the current direction. In the genetic approach, the items are placed at the position where the highest fraction of its perimeter touches edges, either of another item or bin boundaries.

Gonçalves [23] proposed a hybrid Genetic Algorithm based on random keys. The genetic part of this heuristic is responsible to generate and evolve the sequence of items that defines the packing order. The placement method makes use of the Difference Process proposed by Lai and Chan [31] to keep track of the ERS created after each placement.

Bortfeldt and Winter [7] presented a Genetic Algorithm approach to solve the (un)constrained (non-)guillotine SKP and Single large Object Placement Problem (SLOPP, see Wäscher et al. [45]).

Gonçalves and Resende [24] presented a parallel implementation of a multi-population Genetic Algorithm where the chromosome encodes the items sequence and the associated placement rule (bottom-left or left-bottom). The Difference Process is used to keep track of the ERS that are created after placing an item.

Wei et al. [46] presented a Tabu Search using the concept of Skyline representation which is a sequence of line segments that expresses the rectilinear contour of the current packed items.

He et al. [27] presented a two-stage heuristic for solving this problem. In the first phase, a solution is generated placing items in the empty space with the highest fit degree (calculated with the number of touching edges and a smooth degree that considers the number of empty spaces remaining if the item is to be placed at this position), then the solution is improved through a partial tree-search.

Leung et al. [33] presented a heuristic that combines a greedy strategy with a Simulated Annealing algorithm.

Kierkosz and Luczak [30] presented an Evolutionary Algorithm using as a placement algorithm a tree-search that places the items using a bottom-left strategy.

3 Adaptive Sequence-Based Heuristic (ASH)

We propose a multi-start heuristic (see Martí et al. [38]) that iteratively creates a new sequence of items used to define the packing order.

The main concept behind the proposed heuristics is that if a good solution was packed using some ordering (S_{base}), it is possible that a better solution exists changing the order of few items, creating a new packing sequence $S_{current}$. If no improvement is obtained with this new ordering, it may be the case that an ordering with more changes can led to a better solution. So, we incrementally allow more changes to the base ordering. When an ordering generates a new best solution, this will replace the base ordering to be used in the next iterations.

The main steps of the proposed heuristics are shown in Algorithm 1. Starting with S_{base} as the sequence of items ordered by efficiency ($e_i = \frac{v_i}{l_i h_i}$, see Alvarez-Valdés et al. [1]) using v_i as a tiebreaker, at each iteration, a new sequence $S_{current}$ is generated using the current probability α. The sequences are generated (see Algorithm 2) based on the algorithm proposed by Lesh et al. [32] which creates a new sequence adding with a probability of α, one element at a time from the input sequence to the new sequence.

Algorithm 1 ASH main steps

```
S_base ← Items ordered by efficiency with value as tiebreaker
α ← α_min
Generate a new solution with sequence S_base
while stopping criteria are not met do
    S_current ← Generate a new sequence of items based on S_base and α
    Generate a new solution with sequence S_current
    if a new best solution is found then
        S_base ← S_current
        α ← α_min
    else
        α ← min(α + α_inc, α_max)
    end if
end while
```

If the solution generated with $S_{current}$ is the best one so far, this ordering replaces S_{base} and α is set to its minimum value α_{min}. Otherwise, S_{base} is inaltered and α is updated to $\min(\alpha + \alpha_{inc}, \alpha_{max})$.

When a new best solution is found, we seek to intensify the search near to this solution, accomplished setting α to its minimum value to generate sequences that are very similar to base. While higher α allow to diversify the search space producing sequences that differ incrementally more from the base sequence.

Following the ordering of $S_{current}$, the items are packed into the bin, one at a time on the empty rectangle space (ERS) with the smallest area with enough space to fit the item. To prevent the creation of very small empty spaces, after packing an item, we try to pack items that fit exactly into existing ERS.

Algorithm 2 Sequence generator

Input: sequence In, probability α
  $Out \leftarrow \emptyset$
  $n \leftarrow |In|$
  for $i \leftarrow 1, .., n$ **do**
    $j \leftarrow 1$
    $Out_i \leftarrow \emptyset$
    while $Out_i = \emptyset$ **do**
      if $\alpha \leq$ generated random value **then**
        $Out_i \leftarrow In_j$
        $In \leftarrow In \setminus \{In_j\}$
      end if
      $j \leftarrow (j \mod |in|) + 1$
    end while
  end for

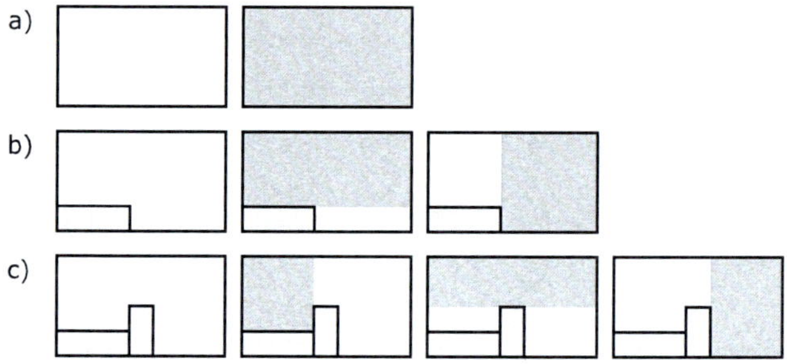

Fig. 1 Difference process

We keep track of the ERS resulting from the packing of items using the Difference Process (Lai and Chan [31]). This process, first, places the box inside the given ERS, then generates the new ERSs that result from the intersection of the box with the existing ERS and removes intersected ERSs. The last phase removes the ERSs that are infinitely thin or are totally inscribed by other ERSs. The Difference Process is illustrated in Fig. 1, in which the darker rectangles depict the available ERS at the beginning of the process (a) and at the end of each item placement (b and c).

The main difference between solving SBSBPP and SKP instances is that for SBSBPP, new patterns are created and added to the solution until all items are packed, while for the SKP, a solution is created considering only one bin.

ASH iterates until a maximum number of iterations has been performed or the optimality is guaranteed (if the solution value is equal to the Continuous Lower Bound (CLB) for the SBSBPP, and if all the items are packed inside the bin for the SKP).

4 Computational Results

The proposed heuristic was implemented in C and the tests were run on a computer with an Intel Core i7-4800MQ at 2.70 GHz with 8 Gb RAM and operating system Linux Ubuntu 18.04.

The next two sub-sections present the computational results for the SBSBPP and SKP, respectively.

4.1 Single Bin Size Bin Packing Problem

We have considered to evaluate the results of our heuristic for solving the SBSBPP, the set of instances, usually referred to as CLASS, generated by Berkey and Wang [5] (CLASS 1 to 6) and by Martello and Vigo [37] (CLASS 7 to 10). This set is divided into 10 classes of 50 instances. Each class contains 5 subsets composed of 10 instances for each value of $m \in [20, 40, 60, 80, 100]$, with $m = \sum_i d_i$. The bin dimensions range from 10×10 to 300×300.

Each instance was run one time generating at most 2000 cutting plans with α_{min}, α_{max} and α_{inc} set to 0.1, 0.65 and 0.002, respectively.

Table 1 presents the results solving all the instances of the dataset CLASS by the heuristics HBP [8], GLS [14], SCH [39], GRASP/VND [41], EA_LGFi [6], and in the last column by ASH. The first line denotes the number of bins needed for the 500 instances, and the average and maximum time to solve a subset. The results from HBP and GLS were retrieved from Boschetti and Mingozzi [8] and those of SCH, GRASP/VND and EA_LGFi were retrieved from Blum and Schmid [6]. The detailed results obtained by ASH solving the SBSBPP are given in Oliveira and Gamboa [40].

Observing Table 1, the best results are clearly obtained by SCH, GRASP/VND and EA_LGFi. ASH could not provide better results than those obtained by the best approaches, but it is extremely effective considering the execution time presented. Noteworthy that the gap between the average and maximum time present by ASH is very tight when compared with the other approaches (e.g., 1.77 s of average time to 27.33 s for the EA_LGFi). This tight gap makes ASH a heuristic with a predictable execution time. Our approach is very fast, simple to implement and can be an effective approach to solve the SBSBPP or to be used in a bounding scheme on more complex solution methods.

Table 1 Summary of the results obtained

Algorithm	HBP	GLS	SCH	GRASP/VND	EA_LGFi	ASH
Number of bins	7275	7284	7243	7241	7239	7269
Average time	21.01	1000	7.89	2.20	1.77	0.04
Maximum time	114.07	1000	55.77	12.00	27.33	0.11

Table 2 Datasets

Dataset	Number of instances	Bin $W \times H$	Items $w \times h$	m
NGCUT	12	$[10-30] \times [10-30]$	$[1-30] \times [1-30]$	$[7-22]$
HADCHR	2	30×30	$[1-22] \times [4-21]$	$[7-15]$
WANG	1	70×40	$[9-33] \times [11-43]$	42
CGCUT	1	40×70	$[9-33] \times [11-43]$	62
OKP	5	100×100	$[1-100] \times [1-99]$	$[30-97]$
LC	3	$400 \times [200-400]$	$[30-200] \times [30-150]$	$[10-20]$
JAKOBS	5	$[65-120] \times [45-80]$	$[5-40] \times [4-36]$	$[20-30]$
LCT	2	$[150-160] \times [110-120]$	$[6-48] \times [4-40]$	$[40-50]$
HT	21	$[20-160] \times [20-240]$	$[1-72] \times [1-113]$	$[16-197]$

4.2 Single Knapsack Problem

To evaluate the performance of our heuristic for the SKP, we have considered the following benchmark datasets, NGCUT (Beasley [3]), HADCHR (Hadjiconstantinou and Christofides [25]), WANG (Wang [44]), CGCUT (Christofides and Whitlock [12]), OKP (Fekete et al. [18]), LC (Lai and Chan [31]), JAKOBS (Jakobs [29]), LCT (Leung et al. [34]), and HT (Hopper and Turton [28]).

The details of the datasets are given in Table 2 which presents, in order of appearance, the dataset name, the number of instances, the range of the bins and items dimensions, and the range of the number of demanded items (m). The first five datasets are referred to as Problems from literature and the remaining four referred to as Zero-waste problems.

The results obtained by ASH are shown in Tables 3 and 4. Each instance was run only once, generating at most 2000 solutions with α_{min}, α_{max} and α_{inc} set to 0.1, 0.9 and 0.005, respectively.

The content of Tables 3 and 4 is the following. The columns identify, in order of appearance, the dataset name, the instance number, the optimal solution, the result obtained by our approach and the result obtained by GRASP [1], Tabu Search [2], and the Parallel Multi-population Genetic Algorithm (MPGA) [24]. For each dataset, the last rows present the number of optimal solutions obtained and the average time needed to solve each instance. Table 4 includes two more rows, the maximum time spent to solve one instance and the average gap between the optimal solution and the solution obtained.

Table 3 Computational results—Problems from literature

Set	#	Z*	GRASP	TABU	MPGA	ASH
NGCUT	1	164	164	164	164	164
	2	230	230	230	230	230
	3	247	247	247	247	247
	4	268	268	268	268	268
	5	358	358	358	358	358
	6	289	289	289	289	289
	7	430	430	430	430	430
	8	834	834	834	834	834
	9	924	924	924	924	924
	10	1452	1452	1452	1452	1452
	11	1688	1688	1688	1688	1688
	12	1865	1865	1865	1865	1865
	Optimum		12	12	12	12
	Avg. time (s)		0.07	0.03	0.01	0.01
HADCHR	3	1178	1178	1178	1178	1178
	11	1270	1270	1270	1270	1270
	Optimum		2	2	2	2
	Avg. time (s)		0.00	0.00	0.01	0.01
WANG		2726	2726	2726	2726	2721
	Optimum		1	1	1	0
	Avg. time (s)		0.77	0.11	0.02	0.01
CGCUT	3	1860	1860	1860	1860	1860
	Optimum		1	1	1	1
	Avg. time (s)		0.39	0.06	0.05	0.02
OKP	1	27718	27589	27718	27718	27486
	2	22502	21976	22502	22502	22119
	3	24019	23743	24019	24019	24019
	4	32893	32893	32893	32893	32893
	5	27923	27923	27923	27923	27923
	Optimum		2	5	5	3
	Avg. time (s)		2.03	1.25	0.16	0.03

Table 3 presents the results obtained by solving the datasets on Problems from literature. The results show that both TABU and MPGA obtained all optimal solutions and that ASH obtained results close to those obtained by the GRASP. Although execution time is not easy to compare, our time remains constantly small on all datasets, while the other approaches it increases greatly as the problem size grows.

The results for the Zero-waste problems are given in Table 4. The results show that ASH obtains a higher number of optimal solutions, and only for one dataset, the average gap is not lower or equal than the one obtained by GRASP.

Table 4 Computational results—Zero-waste problems

Set	#	Z*	GRASP	TABU	MPGA	ASH
LC	1	80000	80000	80000	80000	80000
	2	79000	79000	79000	79000	79000
	3	160000	154600	154600	154600	154600
	Optimum		2	3	2	2
	Avg. time (s)		1.37	0.13	1.26	0.01
	Max. time (s)		4.12	0.38	3.10	0.02
	Avg. GAP (%)		1.13	0	0	1.13
JAKOBS	1	5600	5447	5600	5600	5600
	2	5600	5455	5512	5540	5263
	3	5400	5328	5400	5400	5400
	4	4050	3978	4050	4050	4050
	5	2925	2871	2925	2925	2871
	Optimum		0	4	4	3
	Avg. time (s)		12.68	4.54	4.29	0.02
	Max. time (s)		15.44	16.88	11.52	0.04
	Avg. GAP (%)		2.06	0.21	0.21	1.57
LCT	1	16500	15856	16280	16340	15876
	2	19200	18628	19044	19116	18516
	Optimum		0	0	0	0
	Avg. time (s)		111.39	58.16	19.45	0.05
	Max. time (s)		132.26	63.95	23.71	0.09
	Avg. GAP (%)		3.44	1.07	0.70	3.67
HT	1	400	400	400	400	400
	2	400	386	400	400	386
	3	400	400	400	400	400
	4	600	590	600	600	594
	5	600	597	600	600	600
	6	600	600	600	600	600
	7	1800	1765	1800	1800	1773
	8	1800	1755	1800	1796	1758
	9	1800	1774	1800	1800	1764
	10	3600	3528	3580	3591	3536
	11	3600	3524	3564	3588	3542
	12	3600	3544	3580	3594	3560
	13	5400	5308	5342	5396	5328
	14	5400	5313	5361	5400	5319
	15	5400	5312	5375	5392	5338
	16	9600	9470	9548	9582	9467
	17	9600	9453	9448	9595	9501
	18	9600	9450	9565	9582	9506
	19	38400	37661	38026	38146	37771
	20	38400	37939	38145	38374	38074
	21	38400	37745	37867	38254	37866
	Optimum		3	9	9	4
	Avg. time (s)		612.00	572.30	118.30	0.49
	Max. time (s)		3760.14	5615.75	808.03	2.64
	Avg. GAP (%)		1.50	0.47	0.13	1.24

The average time of our approach maintains very low, while for the three other approaches grows greatly needing in some cases minutes to solve one instance.

Considering the other approaches, ASH is much simpler to implement and to parametrize (i.e., maximum number of iterations, and minimum, maximum and increment of the probability α). GRASP needs a Restricted Candidate List and improvement methods, Tabu Search requires neighbourhood structures and memory strategies, and Genetic Algorithms needs chromosomes and an evolutionary process. The results show that this heuristic, although simple, can generate high-quality solutions using small computing times.

5 Conclusion

We present heuristics to solve the non-guillotine Single Bin Size Bin Packing Problem and Single Knapsack Problem. The heuristics iteratively creates new sequences that define the packing order, incorporating knowledge from the previous one. The computational results show that our approach is competitive with other proposed solution methods for both problems considered. Noteworthy the implementation simplicity and the little parameterization required by this approach. We intend to extend the proposed method to the solution of other cutting and packing problems.

References

1. Alvarez-Valdés, R., Parreño, F., Tamarit, J.M.: A GRASP algorithm for constrained two-dimensional non-guillotine cutting problems. J. Oper. Res. Soc. **56**(4), 414–425 (2005)
2. Alvarez-Valdés, R., Parreño, F., Tamarit, J.M.: A tabu search algorithm for a two-dimensional non-guillotine cutting problem. Eur. J. Oper. Res. **183**(3), 1167–1182 (2007)
3. Beasley, J.E.: An exact two-dimensional non-guillotine cutting tree search procedure. Oper. Res. **36**(1), 49–64 (1985)
4. Beasley, J.E.: A population heuristic for constrained two-dimensional non-guillotine cutting. Eur. J. Oper. Res. **156**(3), 601–627 (2004)
5. Berkey, J.O., Wang, P.Y.: Two-dimensional finite bin-packing algorithms. J. Oper. Res. Soc. **38**(5), 423 (1987)
6. Blum, C., Schmid, V.: Solving the 2D bin packing problem by means of a hybrid evolutionary algorithm. Procedia Comput. Sci. **18**, 899–908 (2013)
7. Bortfeldt, A., Winter, T.: A genetic algorithm for the two-dimensional knapsack problem with rectangular pieces. Int. Trans. Oper. Res. **16**(6), 685–713 (2009). Nov
8. Boschetti, M.A., Mingozzi, A.: The two-dimensional finite bin packing problem. Part I: new lower bounds for the oriented case. Q. J. Belg., Fr. Ital. Oper. Res. Soc. **1**(1), 27–42 (2003)
9. Boschetti, M.A., Mingozzi, A.: The two-dimensional finite bin packing problem. Part II: new lower and upper bounds. Q. J. Belg., Fr. Ital. Oper. Res. Soc. **1**(2), 135–147 (2003)
10. Boschetti, M.A., Mingozzi, A., Hadjiconstantinou, E.: New upper bounds for the two-dimensional orthogonal non-guillotine cutting stock problem. IMA J. Manag. Math. **13**(2), 95–119 (2002)
11. Caprara, A., Monaci, M.: On the two-dimensional Knapsack Problem. Oper. Res. Lett. **32**(1), 5–14 (2004)

12. Christofides, N., Whitlock, C.: An algorithm for two-dimensional cutting problems. Oper. Res. **25**(1), 30–44 (1977)
13. Dolatabadi, M., Lodi, A., Monaci, M.: Exact algorithms for the two-dimensional guillotine knapsack. Comput. Oper. Res. **39**(1), 48–53 (2012)
14. Faroe, O., Pisinger, D., Zachariasen, M.: Guided local search for the three-dimensional bin-packing problem. INFORMS J. Comput. **15**(3), 267–283 (2003)
15. Fekete, S.P., Schepers, J.: On more-dimensional packing I: modeling. Technical report (2000)
16. Fekete, S.P., Schepers, J.: On more-dimensional packing II: bounds. Technical report (2000)
17. Fekete, S.P., Schepers, J.: On more-dimensional packing III: exact algorithms. Technical report (2000)
18. Fekete, S.P., Schepers, J., van der Veen, J.C.: An exact algorithm for higher-dimensional orthogonal packing. Oper. Res. **55**(3), 569–587 (2007)
19. Fleszar, K.: An exact algorithm for the two-dimensional stage-unrestricted guillotine cutting/packing decision problem. INFORMS J. Comput. **28**(4), 703–720 (2016). Nov
20. Furini, F., Malaguti, E., Thomopulos, D.: Modeling two-dimensional guillotine cutting problems via integer programming. INFORMS J. Comput. **28**(4), 736–751 (2016)
21. Garey, M.R., Johnson, D.S.: A Guide to the Theory of NP-Completeness (1979)
22. Gilmore, P.C., Gomory, R.E.: A linear programming approach to the cutting-stock problem. Oper. Res. **9**(6), 849–859 (1961)
23. Gonçalves, J.F.: A hybrid genetic algorithm-heuristic for a two-dimensional orthogonal packing problem. Eur. J. Oper. Res. **183**(3), 1212–1229 (2007)
24. Gonçalves, J.F., Resende, M.G.C.: A parallel multi-population genetic algorithm for a constrained two-dimensional orthogonal packing problem. J. Comb. Optim. **22**(2), 180–201 (2011)
25. Hadjiconstantinou, E., Christofides, N.: An exact algorithm for general, orthogonal, two-dimensional knapsack problems. Eur. J. Oper. Res. **83**(1), 39–56 (1995)
26. Hadjiconstantinou, E., Iori, M.: A hybrid genetic algorithm for the two-dimensional single large object placement problem. Eur. J. Oper. Res. **183**(3), 1150–1166 (2007)
27. He, K., Huang, W., Jin, Y.: An efficient deterministic heuristic for two-dimensional rectangular packing. Comput. Oper. Res. **39**(7), 1355–1363 (2012)
28. Hopper, E., Turton, B.C.H.: An empirical investigation of meta-heuristic and heuristic algorithms for a 2D packing problem. Eur. J. Oper. Res. **128**(1), 34–57 (2001)
29. Jakobs, S.: On genetic algorithms for the packing of polygons. Eur. J. Oper. Res. **88**(1), 165–181 (1996)
30. Kierkosz, I., Luczak, M.: A hybrid evolutionary algorithm for the two-dimensional packing problem. CEJOR **22**(4), 729–753 (2014)
31. Lai, K.K., Chan, J.W.M.: Developing a simulated annealing algorithm for the cutting stock problem. Comput. Ind. Eng. **32**(1), 115–127 (1997)
32. Lesh, N., Marks, J., McMahon, A., Mitzenmacher, M.: New heuristic and interactive approaches to 2D rectangular strip packing. J. Exp. Algorithmics **10**(1), 1.2 (2005)
33. Leung, S.C.H., Zhang, D., Zhou, C., Wu, T.: A hybrid simulated annealing metaheuristic algorithm for the two-dimensional knapsack packing problem. Comput. Oper. Res. **39**(1), 64–73 (2012)
34. Leung, T.W., Chan, C.,K., Troutt, M.D.: Application of a mixed simulated annealing-genetic algorithm heuristic for the two-dimensional orthogonal packing problem. Eur. J. Oper. Res. **145**(3), 530–542 (2003)
35. Leung, T.W.,Yung, C.H., Troutt, M.D.: Applications of genetic search and simulated annealing to the two-dimensional non-guillotine cutting stock problem. Comput. Ind. Eng. **40**(3), 201–214 (2001)
36. Lodi, A., Martello, S., Vigo, D.: Heuristic and metaheuristic approaches for a class of two-dimensional bin packing problems. INFORMS J. Comput. **11**(4), 345–357 (1999)
37. Martello, S., Vigo, D.:. Exact solution of the two-dimensional finite bin packing problem. Manag. Sci. **44**, 388–399 (1998)
38. Martí, R., Resende, M.G.C, Ribeiro, C.C.: Multi-start methods for combinatorial optimization. Eur. J. Oper. Res. **226**(1), 1–8 (2013)

39. Monaci, M., Toth, P.: A set-covering-based heuristic approach for bin-packing problems. INFORMS J. Comput. **18**(1), 71–85 (2006)
40. Oliveira, Ó., Gamboa, D.: Adaptive sequence-based heuristic for the two-dimensional non-guillotine bin packing problem. In: Madureira, A.M., Abraham, A., Gandhi, N., Varela, M.L. (eds.) Hybrid Intelligent Systems, pp. 370–375. Springer International Publishing, Cham (2020)
41. Parreño, F., Alvarez-Valdés, R., Oliveira, J.F., Tamarit, J.M.: A hybrid GRASP/VND algorithm for two- and three-dimensional bin packing. *Ann. Oper. Res.* **179**(1), 203–220 (2010)
42. Scheithauer, G., Terno, J.: Modeling of packing problems. Optimization **28**(1), 63–84 (1993)
43. Singh, K., Jain, L.: Industrial scope of 2D packing problems. Natl. J. Syst. Inf. Technol. **2**(2), 224–237 (2009)
44. Wang, P.Y.: Two algorithms for constrained two-dimensional cutting stock problems. Oper. Res. **31**(3), 573–586 (1983)
45. Wäscher, G., Haußner, H., Schumann, H.: An improved typology of cutting and packing problems. Eur. J. Oper. Res. **183**(3), 1109–1130 (2007)
46. Wei, L., Oon, W.C., Zhu, W., Lim, A.: A skyline heuristic for the 2D rectangular packing and strip packing problems. Eur. J. Oper. Res. **215**(2), 337–346 (2011)
47. Wong, L., Lee, L.S., Serdang, U.P.M.: Heuristic placement routines for two-dimensional bin packing problem. J. Math. Stat. **5**(4), 334–341 (2009)

A CVRP Model for an In-Plant Milk Run System

M. Teresa Pereira, Cristina Lopes, Luís Pinto Ferreira, and Silvana Oliveira

Abstract Automotive industry is very competitive and efficiency is needed in key processes such as logistics in all companies at the supply chain. Using optimization models to increment the efficiency of inbound logistics, such as the in-plant transportation systems, can bring competitive advantages to these companies. This work presents a Vehicle Routing Problem (VRP) model to support and optimize the in-plant transportation system routing, Milk-Run, at the worldwide manufacturing plant of manual assembly in an automotive company, where the routing was planed manually, and too much time and resources were used on this task. At the present, 4 or 5 routes were necessary to supply and collect both raw material and finished items in the plant with 21 workstations. A VRP model was developed into a decision support tool using Integer Programming. The IBM ILOG CPLEX 12.0.8 was used to run the developed model using the production plan of two months of 2018. The results point out an optimal number of 2 milk-runs, a reduction of circa 50%, allowing the freed resources to be used in other production areas of the plant.

Keywords Milk-run system · Vehicle routing problem · Optimization · Case study · Automotive industry

M. Teresa Pereira · L. P. Ferreira
Centre for Research & Development in Mechanical Engineering (CIDEM), Instituto de Ciência e Inovação em Engenharia Mecânica e Engenharia Industrial (INEGI), School of Engineering of Porto (ISEP), Polytechnic of Porto, Rua Dr. António Bernardino de Almeida, 431, 4249-015 Porto, Portugal
e-mail: mtp@isep.ipp.pt

L. P. Ferreira
e-mail: lpf@isep.ipp.pt

C. Lopes (✉)
CEOS.PP/ISCAP/P.Porto and LEMA, Polytechnic of Porto, Porto, Portugal
e-mail: cristinalopes@iscap.ipp.pt

S. Oliveira
School of Engineering of Porto (ISEP), Polytechnic of Porto, Porto, Portugal

S. Relvas et al. (eds.), *Operational Research*, Springer Proceedings in Mathematics & Statistics 374, https://doi.org/10.1007/978-3-030-85476-8_9

1 Introduction

Due to globalization, the supply chain systems have become larger, more complex and more efficiency demanding, especially those working in Lean Thinking. One of the most important operation decisions faced by industrial companies is both the inbound and outbound transportation, either of raw materials and components for manufacturing in a set of workstations, or of goods to a set of customers, using a set of vehicles that travel routes with the minimum possible distance or time.

The distribution of goods concerns the service in a given time period of a set of customers by a set of vehicles, which are located in one or more depots and perform their movements by using an appropriate road network. In particular, the solution of a Vehicle Routing Problem (VRP) calls for the determination of a set of routes, each performed by a single vehicle that starts and ends at its own depot, such that all requirements of the customers are fulfilled, all the operational constraints are satisfied, and the global transportation cost is minimized [8]. When each vehicle has a limited capacity, the problem is called a Capacitated Vehicle Routing Problem (CVRP). It can be applied to real world problems in logistic systems or supply chains, with some modifications for each specific problem [9]. This can also be applied to the in-plant transportation systems. In Lean Production systems, the in-plant milk-run (MR) systems are a common concept for in-plant transportation, also called *mizusumashi*, where materials are delivered from a central storage area (depot) to several points of use (POU) on defined routes and in short intervals (Fig. 1). Furthermore, MR system is a tug system with several trolleys, usually between 1 to 5, that enables frequent deliveries of various materials in smaller lot sizes with short lead times and low inventory, from a central storage area (depot) to various production locations (POU) on collective tours [5].

In a basic MR system, material handlers perform standard routes through a facility at precisely determined time intervals. During each of these intervals, the material handlers pick up containers at a central storage area and, following predetermined standard routes, deliver the material to the POU and return to the central storage area [5]. MR systems provide efficient deliveries in small lot sizes, with a more

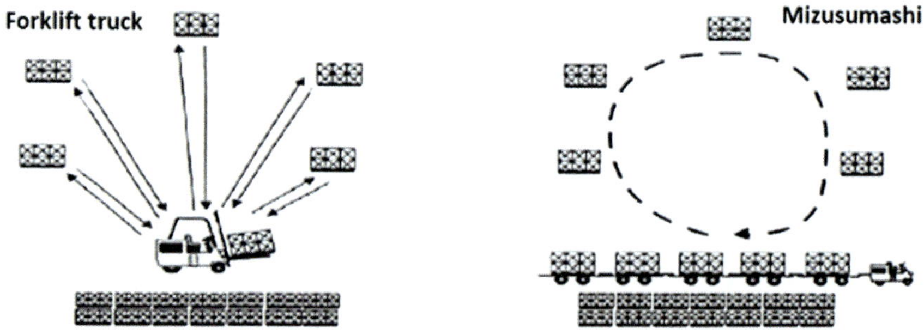

Fig. 1 Internal logistics by operator and by *mizusumashi* [4]

accurate Stock Keeping Unit (SKU) control, allowing a waste reduction, and make the logistic process more organized [1]. Therefore, and since this kind of problems require advanced planning methods to increase the efficiency in logistics, the proposed mathematical model has several mandatory constraints, defined by the company. Being a fixed-course drop-off system, with a defined maximum cycle time per turn, it should guaranty that all POU are supplied, satisfying the demands of the machines. In this case, the objective is to optimize the number of vehicles needed to supply the POU, considering the monthly production scheduling.

This paper is divided into five sections: the first section is an introduction; the second section presents a literature review on the vehicle routing problem (VRP) and its variants; the third section presents a description of the real problem for this study; The fourth section present the mathematical model developed in this study for the real-world problem; the fifth section is the conclusion, which describes some considerations about the work and future work to be developed.

2 Literature Review

Vehicle routing problems are of NP-hard nature [6]. Several researchers have been working with these problems with different approaches concerning exact methods, such as Branch and Bound, Branch-and-Cut, Branch-and-Price, or Column Generation, but which are more time consuming and require more computational capacity if the dimension of the problem is large. Due to this, meta-heuristics such as Genetic Algorithms, Particle Swarm Optimization [6], dedicated heuristics and operating strategies proposed in [5], have had an increasing attention and are extensively used for larger instances of VRP variants.

According to Toth and Vigo [10], a generic definition of VRP can be: given a set of transportation requests and a fleet of vehicles, the problem is then to find a plan to determine a set of vehicle routes to perform all (or some) transportation requests with the given vehicle fleet at a minimum cost; in particular, to decide which vehicle handles which requests in which sequence, so that all vehicle routes can be feasibly executed. In the Capacitated Vehicle Routing Problem (CVRP), the solution calls for the determination of a set of routes that start and end at its own depot, each performed by a single vehicle with limited capacity, such that all requirements of the customers are fulfilled, all the operational constraints are satisfied, and the global transportation cost is minimized [8].

Due to the exiguities of some organizations, the classic problem of vehicle routing has been adapted into many variants. The most common variants of the classic VRP are: the Distance-Constrained and Capacitated VRP (DCVRP), the VRP with Time Windows (VRPTW) [2], the VRP with Backhauls (VRPB), the VRP with Pickup and Delivery (VRPPD), the Open VRP (OVRP), the Multiple Depot VRP (MDVRP), the Mixed Fleet VRP (MFVRP), the Split Delivery VRP (SDVRP) and the Periodic VRP (PVRP). Each can be explored in more detail in [10].

Concerning in-plant MR transportation, with known capacity of trolleys and fixed positions of the POU, a capacitated vehicle routing problem (CVRP) can be applied. Connected to the problem of in-plant transportation with MR, several authors [3, 7] use the Parts Feeding Problem. This concerns on determining the optimum number of MR to be supplied, considering the number of POU in each line and the quantity of material needed, the cycle time of each line, and the supply time to the assembly line [7]. However, in our case the Parts Feeding Problem does not fit due to the differentiation of material quantities existing on the POU of the assembly lines.

Choi and Lee [3] proposed a dynamic supply system for an automobile assembly line based on a heuristic solution. In the first phase, they were based on the expected consumption of material, where the safety stock level on the POU was expected to always be achieved. In another phase, these orders were given to the MR, which dynamically distributed the material to the lines of the factory following the routes formed by the VRP. Taking into account the level of inventory and that the objective is to reduce the number of MR and also to penalize for delivery deviations, the simulations of the algorithm dedicated to the development of the dynamic supply demonstrated to the authors that the number of MR decreased, making the proposed system more efficient than the static supply systems.

According to Klenk et al. [5], one major challenge when designing MR systems is to determine appropriate MR routes and intervals. Individual factory layouts, container demand per POU and inventory ranges need to be considered. This problem bears many characteristics of the well-known VRP. Most MR literature focuses on modelling the MR route-building problem as a VRP variant [5]. Hence, as the problem dimension present in this work is not large, we chose to apply a CVRP to optimize the number of MR concerning the defined constrains and information.

3 Problem Description

Implemented twelve years ago, the milk-run system was an added value to the organization, making it more competitive, functional and structural. The process of supplying the assembly lines is done through the *mizusumashi*, which distributes the various components to the POU. The milk-run system has been adapted to the various types of components transported. The milk-run consists of a traction vehicle (motorcycle) driven by an operator, and a number of towed logistic trolleys (in which the material is transported). The trolleys are divided into three models according to the standards of support for the plates: trolleys for transporting containers, trolleys for transporting polymer boxes, and one trolley for the transport of glass boxes (bonding). The company uses 9 different predefined combinations of these trolleys in the milk-run (Table 1).

The number of boxes of each type is defined considering the maximum weight to avoid damages in the trolley's coupling system which is one of the causes for maintenance problems and increasing cost. Also the number of trolleys is defined to fit the milk run inside the cargo lift and to be possible to move along the narrow

Table 1 The 9 possible configurations of trolleys in the milk run

Trolleys in the Milk Run	A	B	C	D	E	F	G	H	I
Trolleys for transporting containers	1	2	3	0	2	3	1	0	4
Trolleys for polymer boxes	2	1	0	3	2	1	3	4	0
Trolleys for glass boxes	1	1	1	1	0	0	0	0	0

curves that exist on the defined paths. The reduction of the number of milk-runs can reduce costs to the company, regarding contracting, maintenance costs and vehicle purchase.

Customers, in this case, are the 21 POU on the assembly lines that the milk-run will have to supply. Each POU supports a fixed number of assembly machines. The milk-run has to guarantee a supply to the lines, ensuring a constant production. Our focus will be on reducing the number of milk-runs in operation.

In this case study, the POU are grouped horizontally in the plant, so that the milk-run passes in a single corridor and manages to supply the POU at the edge of the line, without having to make detours and sub-courses. As the plant building consists of two floors, each floor has a single main corridor, and the milk-run is able to perform the reversal in the same corridor, thus avoiding unnecessary circuits on the path. After usage of the material in the production cells, the empty carrier material is collected whenever there is a route with available capacity (and never harming the transport of plates), or by another MR dedicated to the end product collection. Therefore the collection is not considered in the model.

4 The Mathematical Model

The model is based on the classic vehicle flow formulation for the VRP [10] which is based on a directed graph $G = (S, A)$, where $S = \{1, 2, \ldots, 21\}$ is the set of POU to supply, and $A = \{(i, j) : i, j \in S, i \neq j\}$ are the arcs representing the shortest path routes connecting the POU. The depot is represented by two nodes 0 (startpoint) and $n + 1$ (endpoint of a route). The set of vehicles is $V = \{1, \ldots, 4\}$, and the set $C = \{1, \ldots, 9\}$ corresponds to the possible configurations of trolleys carried by the milk run. The last five configurations in C can only carry containers or polymer boxes, and the first four trolleys can carry glass boxes Cv.

The parameters are:

t_{ij} is the time (in seconds) of the journey between POU i and POU j,
pb_j is the demand in polymer boxes of POU j,
pc_j is the demand in containers of POU j,
pg_j is the demand in glass boxes of POU j,
cb^o is the capacity in polymer boxes of a vehicle with configuration o,

cc^o is the capacity in containers of a vehicle with configuration o,
cg^o is the capacity in glass boxes of a vehicle with configuration o,
M is an arbitrarily large number.
The decision variables are

$$x_{ij}^{ko} = \begin{cases} 1 \text{ if there is flow in arc } (i,\ j) \text{ using vehicle } k \text{ and configuration } o \\ 0 \text{ otherwise} \end{cases} \tag{1}$$

$$w_j^{ko} = \text{duration (in seconds) of the journey travelled by vehicle } k \text{ with configuration } o \text{ to get to POU } j \tag{2}$$

The objective function (3) minimizes the number of milk-runs.

$$min\ MR = \sum_{k \in V} \sum_{o \in C} \sum_{j \in S} x_{0j}^{ko} \tag{3}$$

$$s.t. \sum_{k \in V} \sum_{o \in C} \sum_{i \in S \cup \{0\}, i \neq j} x_{ij}^{ko} = 1 \qquad \forall j \in S \tag{4}$$

$$\sum_{j \in S} \sum_{o \in C} x_{0j}^{ko} \leq 1 \qquad \forall k \in V \tag{5}$$

$$\sum_{j \in S} x_{0j}^{ko} M \geq \sum_{i \in S, i \neq j} \sum_{j \in S \cup \{n+1\}} x_{ij}^{ko} \qquad \forall o \in C \ \forall k \in V \tag{6}$$

$$\sum_{k \in V} \sum_{o \in Cv} \sum_{j \in S, i \neq j} x_{ij}^{ko} \leq 1 \qquad \forall i \in S \cup \{0\} \tag{7}$$

$$\sum_{i \in S \cup \{0\}} x_{ij}^{ko} = \sum_{h \in S \cup \{n+1\}} x_{jh}^{ko} \qquad \forall j \in S \ \forall o \in C \ \forall k \in V \tag{8}$$

$$\sum_{i \in S \cup \{0\}} \sum_{j \in S, j \neq i} x_{ij}^{ko}\, pc_j \leq cc^o \qquad \forall k \in V \ \forall o \in C \tag{9}$$

$$\sum_{i \in S \cup \{0\}} \sum_{j \in S, j \neq i} x_{ij}^{ko}\, pb_j \leq cb^o \qquad \forall k \in V \ \forall o \in C \tag{10}$$

$$\sum_{i \in S \cup \{0\}} \sum_{j \in S, j \neq i} x_{ij}^{ko}\, pg_j \leq cg^o \qquad \forall k \in V \ \forall o \in C \tag{11}$$

$$\sum_{i \in S \cup \{0\} i \neq j} \sum_{j \in S \cup \{n+1\}} x_{ij}^{ko}\, t_{ij} \leq 1800 \qquad \forall k \in V \ \forall o \in C \tag{12}$$

$$(w_i^{ko} + t_{ij}) - M(1 - x_{ij}^{ko}) \leq w_j^{ko} \qquad \forall i \in S \cup \{0\}, \forall j \in S \cup \{n+1\}, \forall k \in V, \forall o \in C \tag{13}$$

$$x_{ij}^{ko} \in \{0, 1\} \qquad \forall i \in S \cup \{0\}, \forall j \in S \cup \{n+1\}, \forall k \in V, \forall o \in C, i \neq j \tag{14}$$

$$w_i^{ko} \geq 0 \ \forall i \in S \cup \{0, n+1\}, \forall k \in V,\ \forall o \in C \tag{15}$$

The objective function (3) minimizes the number of milk-runs. The first constraints (4) say that all POUs must be visited exactly once in each cycle by one vehicle and one configuration. Constraints (5) assure that every milk-run leaves the depot at most once per cycle. Constraints (6) say that if a vehicle with a configuration visits at least one POU, then it has to leave the depot (node 0).

The constraints (7) force to use only one configuration with glass boxes. Conservation of flow is assured by Eqs. (8) saying that if a vehicle with a certain configuration arrives at a point of use j, it has to proceed in another direction to another point of use h. The set of constraints (10), (9), and (11) limit the delivered quantities to the capacity of the vehicle, in terms of polymer boxes, containers and glass boxes. In Eqs. (12) the cycle time is limited to 1800 s and the constraints (13) prevent from subtours and connect all points of use. Finally, the domain of the decision variables are in (14) and (15).

5 Results

The model was implemented in software IBM ILOG CPLEX Optimization Studio 12.8.0.0.

The optimal number of milk-runs to supply the 21 points of use obtained by the model is only two milk-runs, while the company was currently using four milk-runs for this data. The first milk-run obtained from the model, in Fig. 2, has 1799.52 s of cycle time (30 min), and the second one, in Fig. 3, has 399.49 s of cycle time (6.6 min).

Although the number of milk-runs is minimized, and the desired cycle time is respected, and every POU is supplied, the results produced routes that are not efficient. It is not efficient because the building has two levels and the first route has to change to a different level six times. Each time that it needs to change level, it is mandatory that the milk-run takes a cargo lift which takes a long time, because it is necessary to wait for the lift, then to position the milk-run correctly inside the cargo lift, therefore it can be circa 4–5 min to shift levels.

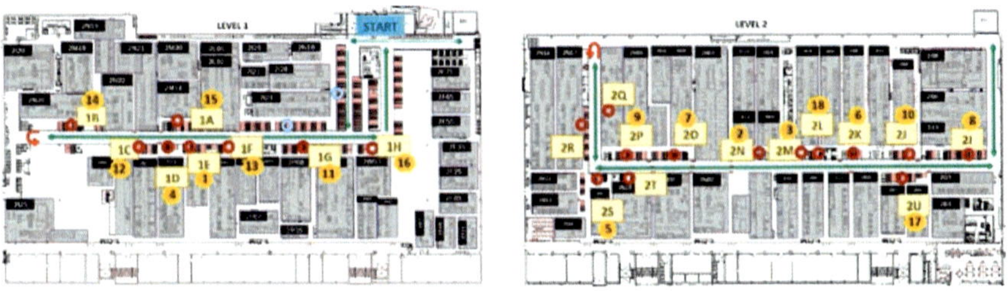

Fig. 2 Route of the first milk-run: DEPOT - 1E - 2N - 2M - 1D - 2S - 2K - 2O - 2I - 2P - 2J - 1G - 1C - 1F - 1B - 1A - 1H - 2U - 2L – DEPOT

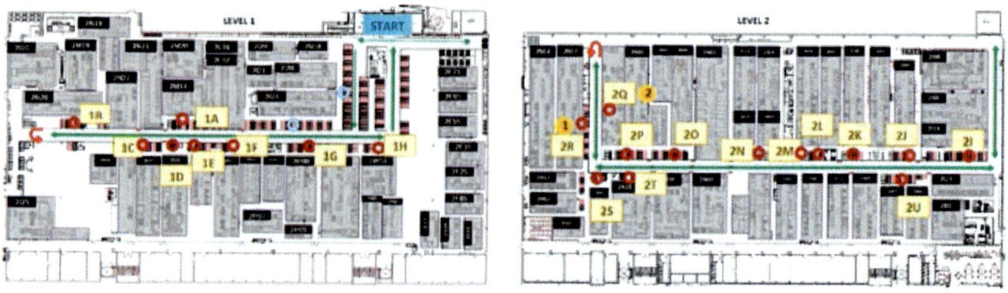

Fig. 3 Route of the second milk-run: DEPOT - 2R - 2Q - DEPOT

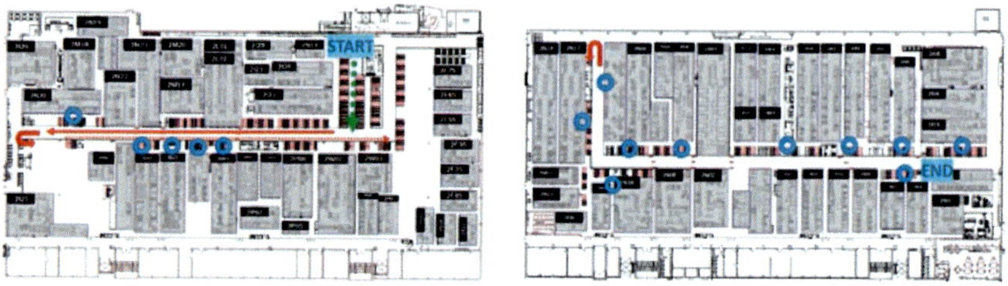

Fig. 4 Route of the first milk-run with the new objective function: DEPOT - 1F - 1E - 1C - 1B - 1D - 2I - 2U - 2J - 2M - 2S - 2R - 2Q - 2P - 2O - 2N - 2L - 2K – END

Therefore, a new objective function TMR was proposed, Eq. (16), which combines the two objectives: minimizing both the cycle time (T) and the number of vehicles (MR). By minimizing the cycle time, the objective function is going to prefer solutions where there have few level changes. The penalty L for using each vehicle was established as 1500.

$$min\ TMR = \sum_{i \in S \cup \{0\}, i \neq j} \sum_{j \in S \cup \{n+1\}} \sum_{k \in V} \sum_{o \in C} x_{ij}^{ko} t_{ij} + L \sum_{j \in S} \sum_{k \in V} \sum_{o \in C} x_{0j}^{ko} \quad (16)$$

With this objective function, the number of milk-runs was optimized to two milk-runs without many level changes in the routes (Figs. 4 and 5). The first milk run takes 851 s from the beginning until returning to the depot, visiting 17 POU, changing levels only once. This milk-run carries 66 units of containers, polymer boxes and glass boxes. The second milk-run has a cycle time of 201 s, visiting only 3 POU, carrying 50 units of only polymer boxes and containers. The solution has the value 4052 for the objective function, giving a total cycle time of 1052 s, with 2 milk runs.

This case study differs from the usual VRP by considering the special kinds of containers in the trolleys and by combining the optimization of the cycle time with the number of vehicles used.

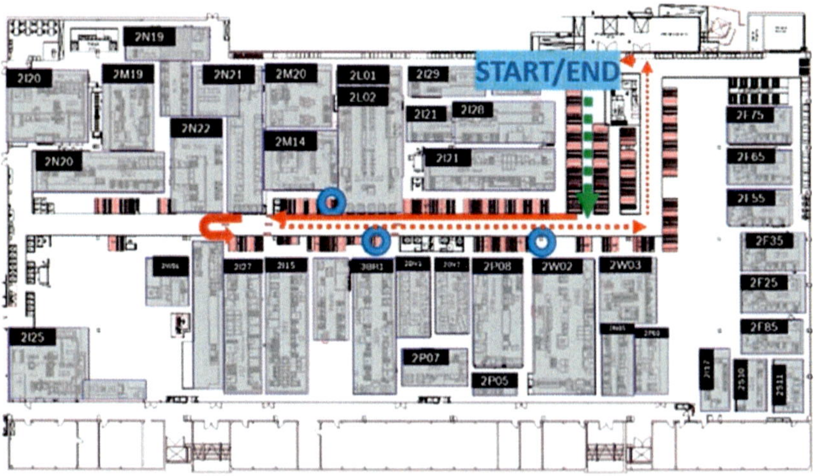

Fig. 5 Route of the second milk-run with the new objective function: DEPOT - 1H - 1G - 1A - DEPOT

6 Conclusions

The developed method was able to present improvements in the number of milk-runs circulating in the building, compared to the current practice in the company. However, although this result was satisfactory, the solutions generated with the first model proved to be not well suited to the real application.

In order to improve the tool, the objective function of the mathematical model was modified, combining vehicle routing with the minimization of cycle time and guaranteeing the minimization of milk-runs, which resulted in more balanced routes. With a penalty imposed on the use of more milk-runs, we obtained a reduction from four to two milk-runs, reducing at least one vehicle than the actual number of vehicles used even during the period of low production, thus demonstrating improvements in relation to the existing supply structure of the company.

In short, it was possible to demonstrate that during the weeks studied, the method optimized the supply circuits, revealing the possibility of reducing costs to the company, regarding contracting, maintenance costs and vehicle purchase.

The decision support tool proves that it is possible to deliver the plates with a reduction of 50% in the number of vehicles and 85.4% in cycle time, in the amount of production studied. With a constant use of the decision support tool in the company, and given that the expected production planning is done several months in advance, it would be possible to predict and eliminate any kind of stoppage associated with milk-run delays, providing for the continuous supply of POU.

Acknowledgements We acknowledge the financial support of CIDEM, R&D unit funded by the FCT – Portuguese Foundation for the Development of Science and Technology, Ministry of Science, Technology and Higher Education, under the Project UID/EMS/0615/2019.

References

1. Arvidsson, N.: The milk run revisited: a load factor paradox with economic and environmental implications for urban freight transport. Transp. Res. Part A: Policy Pract. **51**, 56–62 (2013)
2. Baldacci, R., Mingozzi, A., Roberti, R.: Recent exact algorithms for solving the vehicle routing problem under capacity and time window constraints. Eur. J. Oper. Res. **218**(1), 1–6 (2012). https://doi.org/10.1016/j.ejor.2011.07.037
3. Choi, W., Lee, Y.: A dynamic part-feeding system for an automotive assembly line. Comput. Ind. Eng. **43**(1), 123–134 (2002). https://doi.org/10.1016/S0360-8352(02)00071-2
4. Coimbra, E.: Kaizen in Logistics and Supply Chains. Mc Graw Hill Education, New York (2013)
5. Klenk, E., Galka, S., Günthner, W.A.: Operating strategies for in-plant milk-run systems. IFAC-PapersOnLine **48**(3), 1882–1887 (2015)
6. Kumar, S., Panneerselvam, R.: A survey on the vehicle routing problem and its variants. Intell. Inf. Manag. **4**(3), 66–74 (2012)
7. Kumnuch, N., Prombanpong, S.: Investigation of part feeding problems in manual assembly line. MATEC Web Conf. **26**, 03,011 (2015). https://doi.org/10.1051/matecconf/20152603011.
8. Marinakis, Y., Migdalas, A.: Annotated bibliography in vehicle routing. Oper. Res. **7**(1), 27–46 (2007). https://doi.org/10.1007/BF02941184.
9. Silva, Á., Ferreira, L.P., Pereira, M.T., Neves-Moreira, F.: A model for the multi-depot online vehicle routing problem with soft deadlines. In: International Conference on Innovation, Engineering and Entrepreneurship, pp. 818–824. Springer (2018)
10. Toth, P., Vigo, D. (eds.): Vehicle Routing: Problems, Methods, and Applications, 2nd Edn. No. 18 in MOS-SIAM Series on Optimization. SIAM (2014)

Merging Resilience and Sustainability in Supply Chain Design

João Pires Ribeiro, Bruna Mota, and Ana Paula Barbosa-Póvoa

Abstract In this work, the growing concern with the resilience of organisations and particularly of supply chains (SC) is coupled with the already fundamental environmental factors when designing and planning a SC. A mixed integer linear programming model (MILP) is developed to support strategic/tactical decision making where the cost of emissions is considered in a set of disruptive scenarios under uncertainty. The results contribute to understand the relationship between the two concepts and sustain that coupling these two concepts does not endanger the financial sustainability of the company while increasing resilience and sustainability performance, thus providing a true competitive advantage for those who embrace it appropriately.

Keywords Supply chain management · Resilience · Sustainability · Emissions

1 Introduction

Supply chain (SC) Resilience comes as a recent offspring of Supply Chain Management (SCM), with most of the attention only being given in recent years. Even with a multitude of definitions, one can define such behaviour as follows: "a resilient supply chain should be able to prepare, respond and recover from disturbances and afterwards maintain a positive steady state operation in an acceptable cost and time" [9]. It is a field of study that still has research gaps to explore [6]. One of these research gaps is related to the inclusion of green concerns, where not only economic goals but also environmental concerns are accounted for [4].

J. P. Ribeiro (✉) · B. Mota · A. P. Barbosa-Póvoa
CEG-IST, Instituto Superior Técnico, Universidade de Lisboa, Av. Rovisco Pais, 1049-001 Lisboa, Portugal
e-mail: pires.ribeiro@tecnico.pt

B. Mota
e-mail: bruna.mota@tecnico.ulisboa.pt

A. P. Barbosa-Póvoa
e-mail: apovoa@tecnico.ulisboa.pt

S. Relvas et al. (eds.), *Operational Research*, Springer Proceedings in Mathematics & Statistics 374, https://doi.org/10.1007/978-3-030-85476-8_10

The European Union (EU) has set an objective to reduce greenhouse gas emissions in the EU by at least 40% below 1990 levels by 2030 [3]. To achieve this goal one of the policy frameworks underway is the European Union Emissions Trading System (EU ETS). Through this framework, also known as the "cap-and-trade" system, a maximum limit is imposed on the amount of greenhouse gases emitted. This limit is reduced over time so that total emissions are continuously decreasing. Within this set, organisations can buy emission allowances, which as the name indicates allows for a certain amount of CO_2 emissions. If the organisation is able to stay under the estimated emissions level, it can sell and trade these allowances or keep them for future needs. Allowances have become commodities and auctioning is the default method for allocating them. Its value has been rising over the last year, currently reaching +153% of the reference value that existed until mid-2018, showing the utmost importance of considering CO_2 costs in the supply chain decision-making process [2].

It is then very relevant to explore the coexistence of environmental and resilience concerns when designing and planning a SC [1, 5], considering the two growing topics of SC are rarely considered in combination. This study comes in a time of great pertinence to practitioners that are now making long-standing decisions in their SC, with the changes that come from new trends that complement the sustainability as is the case of adaptability to new markets trends at a very fast pace. This type of decisions tends to be complex, with significant investments to be implemented in the long term, adding special necessity for decision aiding tools regarding strategic and tactical decisions.

This paper addresses this need and develops an optimisation model that explores the relationship between Resilience and Green goals when designing and planning a SC.

A case study of a European SC is provided, applying a new model where resilience and environmental concerns are considered, apart from the common SC profit. Managerial insights are obtained to help in the process of acknowledging the full impacts of disruptive events on the fitness of operations under different scenarios. Furthermore, attention is brought to how much proactive management can contribute to a competitive advantage.

2 Problem Characteristics

The SC to be modelled is generic and involves five different echelons: Raw Materials Suppliers, Plants, Warehouses, Markets and Outsourcing. The possibility to outsource the production of final products, eases the production side of the SC and those products are shipped directly to markets (see Fig. 1). It allows for direct flows between consecutive SC echelons, transhipment in plants and warehouses, direct flows between warehouses and markets and it also includes the possibility of reverse flows. Three types of reverse flows are considered, namely for:

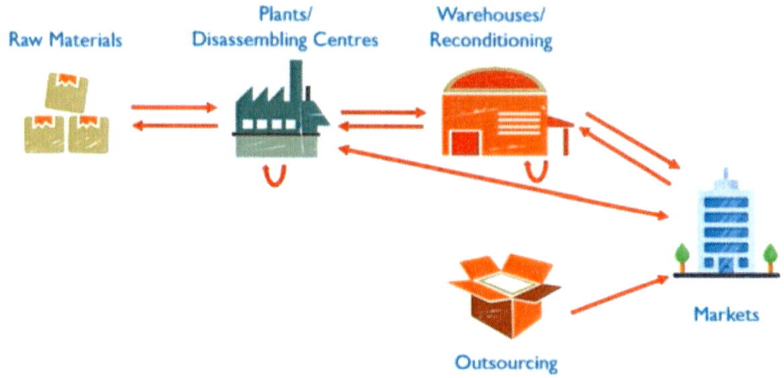

Fig. 1 Schematics of the supply chain structure

- Non-conforming products collected from the markets that go to warehouses to be reconditioned, repacked and introduced in the forward SC;
- End-of-life products that are collected, when the expected lifetime of the product is reached, and are refurbished to be introduced in the forward SC;
- Products that fail to be refurbished are sent to landfill.

3 Model Description

When studying resilience there is the need to define disruptions on which to test the model, an activity that can prove itself challenging. One must define disruptions that represent a broad set of possible disruptions without falling in the error of defining disruptions that are too specific thus failing to meet the need for managerial insights in the context of SC Resilience. Disruptive events are associated with some kind of unknowns in terms of risks, which leads to not being possible to always comprehensively identify the disruptive event and its consequences. By choosing a type of disruptive events it is possible to generate a representative set of disruptions that meet the objective of providing generic, but useful, results. These disruptions are then studied on each run of the model, as will be detailed below.

The environmental assessment is explored using a Life Cycle Assessment (LCA) approach provided by specialised software. In this case, SimaPro software was used to determine the needed parameters, namely:

- CO_2 emissions from truck transportation in forward flows "Forward impact" (per kg.km)
- CO_2 emissions from truck transportation in reverse flows "Reverse impact" (per kg.km)
- CO_2 emissions from final product production "$Production_p$" (per kg/product).

These parameters in combination with the cost of 1 Kg of CO_2 are used in the model so as to quantify the environmental impact of the SC.

The model possesses relevant characteristics as it includes a closed loop SC, with not only forward logistics but also reverse logistics and entails uncertainty, relevant when considering the time span needed for tactical and strategic decisions. Uncertainty is considered in the SC demand and on disruptions. A scenario tree is constructed combining these two sources of uncertainty. Demand variability is introduced by generating a scenario on each period from a set of three possibilities (Pessimistic, Realistic or Optimistic). The probabilistic nodes are then combined with the variability from the disruption that can only assume two options, or it occurs, or it does not occur in a specific time period to consider. This case study involves a 15 year time span, instantiated in three five years periods. With this, each scenario probability is given by the probabilities of the path, with all stages, between the root node and each final leaf node.

The proposed model relies on the maximisation of Expected Net Present Value (ENPV), also considering environmental costs, under a set of disruptions and under uncertainty.

$$\max \text{ENPV} = \sum pb_s \times NPV_s \tag{1}$$

Expected Net Present Value is calculated considering each scenario, in the scenario tree, using cash flows and an interest rate. The net earnings come as result of the difference between the income (Sales revenue) and the costs (the refunds from non-conforming products, purchasing products, operational, inventory, transportation, depreciation of invested capital considering taxes and also the environmental costs defined in Eqs. 2 and 3.

Two environmental cost functions are considered, one to account for transportation (Eq. 2) and another to account for the production of final products (Eq. 3).

The costs of environmental impact of transport are given by the summation of emissions of transport of production in both forward ($ForwardFlow_{v,w,p,s,t}$) and reverse flows ($ReverseFlow_{v,w,p,s,t}$) times the distance between entities ($Dist_{v,w}$) multiplied by the cost of emitting one kg of CO_2 ($Forwardimpact$ or $Reverseimpact$), as shown in Eq. 2:

$$\begin{aligned} &\text{Environmental cost of transport} = Cost_{CO_2} \times \\ &\left(\sum_{v,w,p,s,t} ForwardFlow_{v,w,p,s,t} \times Dist_{v,w} \times \text{Forward impact} \right. \\ &\left. + \sum_{v,w,p,s,t} ReverseFlow_{v,w,p,s,t} \times Dist_{v,w} \times \text{Reverse impact} \right) \end{aligned} \tag{2}$$

The environmental impact of production ($production(p)$) is given by the summation of emissions when a specific product p is produced multiplied by the cost of emitting one kg of CO_2 ($Production_p$), that is given by Eq. 3. Emissions for reverse product flows are expected to be lower than those that are produced directly from raw materials.

$$\text{Environmental cost of production(p)} = Cost_{CO_2} \times \left(\sum_{v,w,s,t} ForwardFlow_{v,w,p,s,t} \times Production_p + \sum_{v,w,s,t} ReverseFlow_{v,w,p,s,t} \times Production_p \right) \quad (3)$$

Overall the model is subject to entity capacity constraints, transportation constraints and technology constraints so as to correctly and thoroughly model the SC.

4 Case Study

The model is applied to the Case Study presented by [7], an European SC that is interested in the improvement of its network. The initial network relies on one plant in Hamburg with the possibility of opening new plants in Bilbao and/or in Milan. Each one of the new plants has a set of suppliers associated. There is also the possibility to open new warehouses or to upgrade the already existing ones. The activities performed by each entity are categorised as production technologies, that result in intermediate products (P16 to P27); and assembling technologies, that combine products and result in the final products (P29 to P34). Reverse logistics is implemented by disassembling technologies that can use final products (as defined in the model description) to generate intermediate products. The tactical and strategic decisions involved are all free, that is that in each time period the decision maker can change the design of the network considering the inherent costs.

In this work, the SC will be tested under a reference scenario, where no disruption occurs, and three types of disruptions as defined by [10] which can represent an realistic set of possible types of disruptions:

- **Disruption 1:** Supply—100% decrease in the production capacity of the most important plant in time period 2
- **Disruption 2:** Production—The most important raw material suppliers have their supply suspended in time period 2
- **Disruption 3:** Transportation—The flows that carry the highest quantity of products in time period 1 are stopped in time period 2.

5 Results

The developed model was applied to the case-study and two different approaches were compared: one with an objective function considering strictly economic concerns (Table 1) and another with the new objective function where environmental impact is translated into costs and considered in the optimisation objective (Table 2). The results can be summarised in four outputs: the objective function value (ENPV);

Table 1 Economic model results

	ENPV (€)	ECSL	Investment (€)	Complexity
Reference	2,05E+07	99,40%	7,91E+06	105
Disruption 1	2,04E+07	99,00%	8,06E+06	110
Disruption 2	1,99E+07	96,20%	8,97E+06	112
Disruption 3	2,00E+07	99,30%	8,36E+06	122

Table 2 Environmental model results

	ENPV (€)	ECSL	Investment (€)	Complexity
Reference	2,01E+07	99,998%	8,01E+05	106
Disruption 1	1,86E+07	99,994%	1,01E+06	108
Disruption 2	1,87E+07	99,989%	1,12E+06	109
Disruption 3	1,92E+07	99,998%	8,49E+05	107

the Expected Customer Service Level (ECSL), that is the difference between the demand and the flows that reach the markets; the necessary investment (includes: initial investment for the existence of that entity, the investment in allowing the use of that entity, any expansions made to any SC entity, warehousing investment costs); and the quantity of flows present in the SC network design (Complexity). Each of these outputs is depicted for the reference case (where no disruption occurs) and for the cases where disruptions occur.

Comparing the two approaches one can see that there is a clear reduction in ENPV when considering environmental impacts, however without endangering the economic sustainability of the company. This reduction is expected as the news costs with CO_2 emissions being now accounted for. When analysing the disruptions, it can be seen that the relative impact of the different disruptions changes. While disruption 2 is the one that produces the higher impact in the ENPV and ECSL (lower values) and investment (higher value) when only strictly economic concerns are at stake (see Table 1), when environmental impacts are considered disruption 1 is the one with higher impacts (see Table 2). However, in both cases disruption 2 is the one that affects the service level the most (with lower impact in the case when environmental concerns are considered). Taking into consideration environmental concerns is observed an impact on ENPV, due to the increased costs of transportation. This change leads to a decrease in own production and an increase in outsourcing so as to reach the best value of sales and consequently the best return with such SC.

Additionally, in terms of investment, it can be seen that accounting for environmental costs (Table 2) leads to lower levels of investment when responding to disruptions occurrence. This behaviour comes from the reduction in the SC complexity that reduces the investment needed to open new locations and contract fees from opened network connections. Note however that these results are associated with lower financial returns, although still not compromising SC profitability. The

Table 3 Economic model—variation in results between the disruptions scenarios and no disruption (%)

	ENPV (%)	Investment (%)
Disruption 1	0	2
Disruption 2	−3	13
Disruption 3	−3	6

decrease in complexity is explained by the fact that if the complexity increases then it increases the flows occurrence and consequently with more flows, higher environmental cost in addition to the increased costs in installing new facilities. Therefore, when environmental impacts are considered it is preferred to focus the activities in fewer locations. This is applied, and perceived in the values of SC Flow Complexity, by delaying the installation of new facilities until they are essential for the SC success. For example, the installation of the reprocessing facilities are delayed in the environmental model so as to reduce emissions costs. While in the economic model the reprocessing facilities (four entities) are implemented in the first period, in the environmental model it is not like so. When no disruption occurs, reprocessing facilities are only implemented in the second period. When disruptions occur, one (in disruptions 1 and 2) or two (in disruption 3) reprocessing facilities are open in the first period, so as to find a balance between the alternative revenue and the increased cost in emissions when a part of the SC is not fit for business.

In terms of ECSL, when environmental costs are considered, the SC retains high levels of service quality, even when disruptions occur and comparing to the results obtained with strictly economic concerns. This positive result of meeting customer demand is, however, the result of high use of outsourcing. In fact, although emissions of the final product are considered, the emissions related to the upstream transport are not, thus bypassing the added environmental cost of transportation. Additionally, the studied SC is able to almost fulfil the plenitude of demand, with the difference between cases being minimum.

Analysing now the variation in results between the disruptions scenarios and no disruption in the economic model (Table 3), one can see that ENPV relative change is generically lower than that in the model with environmental concerns (Table 4), coming from the SC ability to less restrictively adapt to the new conditions. The absolute values in the environmental model (Table 2) are lower than those of the economic model (Table 1), much due to the increased pressure to meet sustainable goals while meeting customer demand which leads to a higher usage of outsourcing.

Analysing now the case when environmental concerns are considered it can be seen that a positive relationship exists between resilient behaviour and the increased flow complexity, there is an increase in SC flow to face the disruptions—a fact that is aligned with previous work when strictly economic concerns are considered [7, 8]. The disruptions in operations lead to the need for increasing the flexibility in operations. The multiple options introduced by having more flows are used as a mean

Table 4 Environmental model—variation in results between the disruptions scenarios and no disruption (%)

	ENPV (%)	Investment (%)
Disruption 1	−8	26
Disruption 2	−7	40
Disruption 3	−5	6

Table 5 Environmental costs (€)

	Transportation	Production
Reference	6,62E+04	4,50E+05
Disruption 1	7,44E+04	4,63E+05
Disruption 2	6,79E+04	4,52E+05
Disruption 3	7,57E+04	4,40E+05

to compensate the disruptions imposed, not compromising customer satisfaction. Due to the nature of this model, one can add that having a more flexible and responsive SC can help companies to operate a sustainable, resilient SC.

The environmental costs obtained are shown in Table 5. One can see that in the reference case there is a lower cost associated with transportation emissions than when disruptions occur. Disruption 2, a production disruption, has the lowest impact of the three disruptions since the lower production reduces the amount of goods available for shipment thus reducing the amount of goods transported. Disruption 3, a transportation disruption, generates the need to compensate the flows that are disrupted by having longer routes thus higher emissions, which ultimately leads to the worst environmental impact.

The production impacts are more homogeneous between the different cases. Disruption 1, a supply disruption, is responsible for the worst impact since the lack of raw materials leads to the creation of more inventory no matter the individual environmental characteristic of each product.

From the above results, it is possible to present a robust comparison of different approaches to resilience and add on the SC resilient behaviour when considering environmental concerns, concluding on the importance of quantifying environmental concerns when addressing resilience in SC.

6 Conclusion and Future Work

Acknowledging the importance and how can one apply the relationship between SC Resilience and SC Sustainability is of extreme relevance.

In this work, a MILP model that incorporates resilience and sustainability concerns is presented. The model is applied to a case study where three different disruptions types are implemented. Sustainability concerns are taking into consideration by considering the costs defined by European Union Emissions Trading System (EU ETS).

Our results, generically show that designing a SC with environmental concerns, applying such impact in terms of cost in the objective function, does not endanger the profitability of the company nor the ability to be resilient. The main differences rely on more efficiently managed flows and the production of products based on its environmental performance (transportation and production).

When designing a resilient environmentally conscious SC a lower investment level is obtained, when compared to a network design for a pure economic model. The added flexibility and redundancy are contributing factors for higher emissions thus a leaner SC is preferred when aiming for a green objective. However, this leaner and cheaper SC does not provide the same economic return. In fact, the ENPV decreases, slightly, when comparing the results from the two models, while maintaining a very good customer service level. This decrease in economic return is, in the case studied, not drastic as it does not endanger profitability and economic sustainability for the company.

Thus, designing and planning a SC taking into consideration resilience and environmental concerns, takes advantage of the increased flow complexity to face the SC disruptions. The increased complexity, balanced by the environmental cost, adds flexibility and redundancy to keep the SC operational.

These results empower decision makers to pursue this type of strategies performing a trade-off between return, investment and environmental impact.

- Future work should include the application of the model to different case studies so as to increase understanding on the relationship between SC Resilience and SC Sustainability. This should mean applying the model to different industries and to different geographical reach.
- The integration of the operational and tactical decision level of SCM would also be important to explore since it involves different strategies. In fact, a scope where studying speed, in the context of SC Resilience, could be possible.

Acknowledgements The authors acknowledge the support provided by FCT and P2020 under the project PTDC/EGEOGE/28071/2017, Lisboa -01.0145-Feder-28071.

References

1. Barbosa-Póvoa, A.P., da Silva, C, Carvalho, A.: Opportunities and challenges in sustainable supply chain: an operations research perspective. Eur. J. Oper. Res. (2018). https://doi.org/10.1016/j.ejor.2017.10.036. http://www.sciencedirect.com/science/article/pii/S0377221717309499
2. European Energy Exchange AG: CO_2 European emission allowances (2019). https://markets.businessinsider.com/commodities/co2-emissionsrechte
3. European Union: 2030 climate & energy framework (2019). https://ec.europa.eu/clima/policies/strategies/2030

4. Fahimnia, B., Jabbarzadeh, A.: Marrying supply chain sustainability and resilience: a match made in heaven. Transp. Res. Part E: Logist. Transp. Rev. **91**, 306–324 (2016)
5. Jabbarzadeh, A., Fahimnia, B., Sabouhi, F.: Resilient and sustainable supply chain design: sustainability analysis under disruption risks. Int. J. Prod. Res. **56**(17),5945–5968 (2018). https://doi.org/10.1080/00207543.2018.1461950
6. Kamalahmadi, M., Parast, M.M.: A review of the literature on the principles of enterprise and supply chain resilience: major findings and directions for future research. Int. J. Prod. Econ. **171**, 116–133 (2016)
7. Ribeiro, J.P., Barbosa-Póvoa, A.: Supply chain resilience: an optimisation model to identify the relative importance of SC disturbances. In: Operational Research. IO 2018, vol. 278, chap 13, pp. 189–198. Springer, Cham (2019). https://doi.org/10.1007/978-3-030-10731-4-13
8. Ribeiro, J.P., Barbosa-Póvoa, A.: Modelling and analysing supply chain resilience flow complexity. Comput. Aided Chem. Eng. **43**(28th European Symposium on Computer Aided Process Engineering, 815–820 (2018). https://doi.org/10.1016/B978-0-444-64235-6.50143-1, https://linkinghub.elsevier.com/retrieve/pii/B9780444642356501431
9. Ribeiro, J.P., Barbosa-Póvoa, A.: Supply chain resilience: definitions and quantitative modelling approaches - a literature review. Comput. Ind. Eng. **115**, 109–122 (2018)
10. Rice, J.B., Caniato, F.: Building a secure and resilient supply network. Supply Chain. Manag. Rev. **7**(5), 22–30 (2003)

Analysis and Optimisation of a Production Line Using Discrete Simulation

Fabiane K. Setti, Carla A. S. Geraldes, João P. Almeida, and Marcelo G. Trentin

Abstract This study presents a simulation-based procedure to analyse a production line of a metalworking company. We use a simulation tool, ProModel® software, to reproduce the existing production line layout of the company best-selling product which represents about 70% of the total sales. Our purpose is to get information about the existing system behaviour, and to find strategies to increase actual production level to meet the market's demand. Based on an initial simulation model, different production scenarios were proposed and results have shown that it is possible to increase the production level allowing to meet the increasing demand for the product. The following changes in the production system were considered: (i) the use of intermediate stock of work-in-process items, (ii) the introduction of new equipment, and (iii) a mixed strategy where the introduction of new equipment is combined with the use of intermediate stock of work-in-process items. In summary, this research exhibits the flexibility of the simulation technique to address manufacturing problems throughout the creation of different scenarios providing some of the behaviour of the systems allowing the anticipation of final outputs.

Keywords Discrete simulation · Manufacturing process · Bottleneck

---

F. K. Setti
Instituto Politécnico de Bragança, Campus de Santa Apolónia, 5300-253 Bragança, Portugal
e-mail: fabisetti@hotmail.com

F. K. Setti · M. G. Trentin
Universidade Tecnológica Federal do Paraná, Pato Branco, Paraná, Brazil
e-mail: marcelo@utfpr.edu.br

C. A. S. Geraldes (✉) · J. P. Almeida
CeDRI – Research Centre in Digitalization and Intelligent Robotics, Instituto Politécnico de Bragança, Campus de Santa Apolónia, 5300-253 Bragança, Portugal
e-mail: carlag@ipb.pt

J. P. Almeida
e-mail: jpa@ipb.pt

C. A. S. Geraldes
Centro ALGORITMI, Universidade do Minho, Campus de Azurém, 4800-058 Guimarães, Portugal

S. Relvas et al. (eds.), *Operational Research*, Springer Proceedings in Mathematics & Statistics 374, https://doi.org/10.1007/978-3-030-85476-8_11

1 Introduction

Manufacturing companies facing increasing demands are constantly challenged to be innovative and to produce their products at reduced lead-times [13]. Strong market competition fosters producers to identify their weak points of the manufacturing processes and to implement changes. Hence, optimising both manufacturing processes and production lines are key issues that nowadays manufacturing companies face in order to meet increasingly demanding market requirements.

According to Liu and Chiang [3, 10], bottlenecks may exist in production environments and, in this case, they are identified as the main factors that affect operations management and the efficiency of production lines. This can be due, for example, to the fact that the capacity of the production line is insufficient to meet the current demand.

Simulation is a powerful technique widely used in decision-making processes for engineers, designers and managers. This tool enables the analysis and evaluation of a wide range of scenarios that would not be possible otherwise [16].

It is possible to define computer simulation as a method that imitates the operation of a system and its internal processes, usually over time, and in appropriate detail to comprehend and draw conclusions about the system's behaviour [17]. In several cases, simulation method also includes testing and comparing alternative scenarios [18].

The simulation model to be created should be as similar as possible to the real system. For this purpose it is important to identify some aspects, such as the definition of the problem to be solved and the collection of all the information and data required to build the model. As for the model itself it is important to choose the type, evaluate it, build it in some computer language, validate it, verify if it is sensitive to changes concerning the input parameters, and also if the results are consistent and representative [7].

Simulation models can be applied in many different fields like, for instance, airports, hospitals, economic systems, restaurants, ports, supply chains, manufacturing, military, telecommunications, public sector, customer services, emergency-response systems, among others [9, 17]. Because of this versatility, many studies are being developed using a simulation software available on the market.

In this work, we present the development of a simulation model using ProModel® software. We present the case study of a metalworking production company located in South region of Brazil. The company produces a wide range of products (pans), however, its best-selling product, by itself, is responsible for about 70% of total sales. The aim is to discuss the main issues concerning the principal production line of the metalworking production company. The main purpose is to investigate new strategies and procedures to increase the production level and identify the bottleneck in the production flow.

The present work is organised as follows: the next section presents a short literature review about simulation in the manufacturing context. Afterwards, in Sect. 3 it is described the layout production line under study. In Sect. 4 we describe the methodology leading to the development of the simulation models for each considered scenario. Simulation results and related discussion are shown in Sect. 5 and, finally, conclusions are presented in Sect. 6.

2 Simulation

Simulation can be understood as a technique to perform experiments using models that describe the behaviour of a system during a period of time [8]. According to Mourtzis, Doukas and Bernidaki [13] simulation comprises several technological tools and methods that can be implemented on manufacturing due to its experimentation and validation of product, process and system configuration.

The utilisation of simulation is increasing by virtue of its many benefits. As a result of that managers are integrating simulation in their regular operations on an increasingly basis [1].

It is possible to define simulation optimisation as a process that aims to find the best input variable values between all the possibilities. The main objective of simulation optimisation is to minimise the resources consumed and at the same time maximise the information obtained in a simulation experiment [2].

Müller et al. [14] used the simulation software Arena® for the improvement of a productive real flow. The model was created according to the production phases and subsequently two proposals of change were considered. One concerning the addition of new equipment and the other reallocating the existent resources. Both scenarios were simulated and compared with the real system. Results showed that it was possible to increase the productivity of the production line in both scenarios. In particular, in one of the scenarios productivity increased without the need to purchase a new equipment.

Hernández and Librantz [6] used ProModel® software to study different strategies on export commodities logistic chain. With the aid of a simulation model it was possible to verify the main parameters and variables present in the system. In addition the critical points concerning the absence of operation time were also identified.

Hassan and Abdelsalam [5] worked in a simulation study using SIMIO® software to analyse the behaviour of a supply chain of the contraception methods for Egypt Ministry of Health (MOH). This was a pioneer study using simulation on this specific field in Egypt.

Sanjaya and Perdana [15] developed a study using simulation to improve a logistic system in Pangalengan, Indonesia. In their work four scenarios were created aiming the increase of quality, continuity of production and cost reduction. Their best scenario found has lower costs than the current model.

Loaiza, Sarmiento and Correa [11] developed a study using the simulation ProModel® software to increase production of an existing system. The productivity of a flexible manufacturing cell was evaluated, in order to measure and choose the best alternative to maximise the use of the available resources.

Nowadays there are a great amount of other studies that benefit of the simulation's versatility. In the next section we approach a case study that aims to optimise a production line through discrete simulation, using real data.

3 Case Study

This work involves the analysis of a case study, which incorporates modelling and computer discrete simulation into the identification of production bottleneck and find strategies to face market's demand. The company involved in this work produces pans and it is located in the State of Paraná, Brazil. During the data collection period of time, the company had 16 employees assigned to the manufacturing area and consumed, in its production, about ten tons of aluminium per month.

Although the company manufactures a wide range of products, 70% of its total production level is concentrated on a specific product. For this reason, and also because the company is facing problems to meet the existing demand, the production line of the main product was chosen to be modelled and analysed. All the input data for this product were determined based on the available technological data of the process and on the data concerning the material flow during production.

Figure 1 illustrates the flowchart of the analysed production line. Production line starts with a dedicated storage area for the raw material. In this area, a specific amount of aluminium is positioned in this waiting area to be manufactured. The used raw material consists in circular aluminium sheets that need to be processed in seven different workstations (machines) to reach the final shape of the product.

The raw material passes through a press machine. In this workstation the product acquires the predefined depth of the final product. Afterwards, the work-in-process (WIP) item passes to a cleaning process that is manually performed. This step is followed by the cutting and bending machine that is responsible to finish the superior part of the pan, making an internal bending and an oval cut. The WIP item is then introduced in an industrial sanding machine to remove the existent imperfections, and later the finishing process is performed by an industrial polishing machine. The next stage in the production line includes a riveting machine. At this workstation rivets sets are done to later join the plastic handles. The last workstation in the production line is responsible for the assembly process. At this workplace the plastic handles are fixed to the body of the pans with the rivets.

The raw material to feed the production line arrives once a day. For this reason our analysis was conducted during a normal work shift (7:30–18:00). The simulation

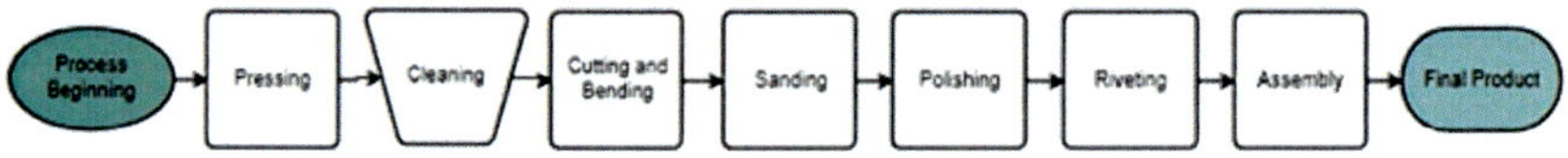

Fig. 1 Production Line Flowchart

period also takes into account three broken intervals: a lunch break of 1H30 min, a 5 min stop to exchange sandpaper, and 10 min stop to perform a scheduled cleaning operation. Removing all the pauses, the working hours available for the production line are 8H45 min. This was the time considered in the simulation model.

4 Methodology

In this study we present a simulation-based procedure with the purpose of analysing a current production system, and also evaluate different production strategies in a production line. Simulation technique was chosen rather than other methods (such as: Theory of Constraints or Assembly Line Balancing), since the focus of this work was not only to find the bottleneck on the production line, but also to develop scenarios where the daily production batch was totally produced. For that specific purpose simulation technique was considered a more multipurpose tool.

Our aim is to analyse the existing production system behaviour and identify new production strategies. For that purpose, a simulation model was developed and different scenarios were analysed, such as: (i) scenarios that use intermediate stock of WIP products, (ii) scenarios with the addition of new equipment, and (iii) a mixed strategy that combines the addition of new equipment and intermediate stock of WIP.

ProModel® software student version choice was dictated by the availability of the software during the research period. Nevertheless, the selected tool imposed some limitations. The most significant one was related with the number of locations that could be used in our model. The maximum number of locations was limited to 20 and for that reason all the considered scenarios were developed in a way to not exceed the limitations of the software version available.

The developed simulation model corresponds to the actual production line (see Fig. 2). The arrival area included in the model represents the storage area of the line raw material. Along the production process, the WIP products flow through all the production line workstations. That includes a press machine, the cleaning area, the cutting and bending equipment, an industrial sanding machine and an industrial polishing machine, riveting machine and lastly the assembly process workplace.

All the input data (i.e. processing times for each of the seven workstations) were provided by the company. From previous studies it was possible to characterise the processing time of the line workstations. For this reason, data were accepted as representative of the reality of the factory. Table 1 presents the processing times, in seconds, used in the simulation model. All the considered times were statistically modelled by means of an uniform distribution.

Both the base model and the different scenarios were based on a production period of one day (8.75 h of simulation time) and 23 replications were programmed, considering an average of twenty three working days during a month.

For model validation purposes, the simulation of the base model was performed during a working day, and five replications were programmed, representing the working days during a normal week. The daily production level reached using our sim-

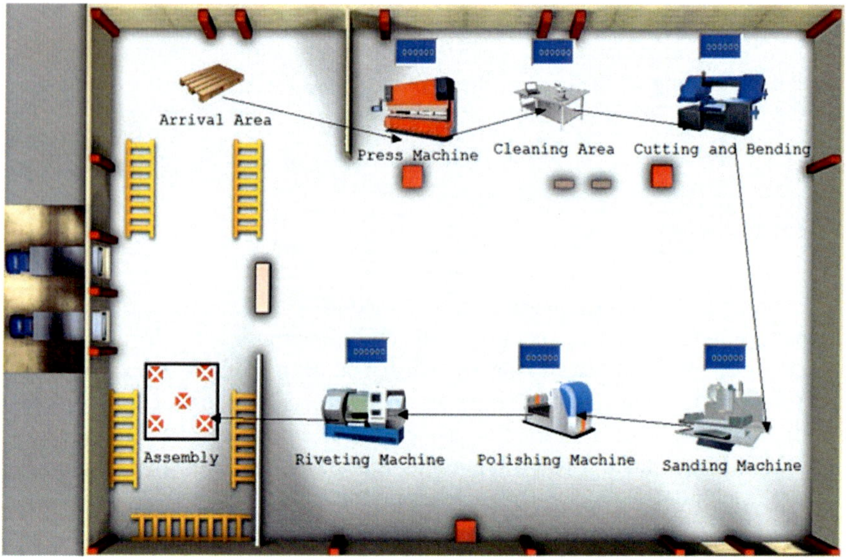

Fig. 2 Computer model of the existing production system in ProModel® software environment

Table 1 Processing time of the workstations

Workstation	Time (seconds)
Press machine	U (18, 19)
Cleaning area	U (20, 24)
Cutting and bending machine	U (70, 74)
Sanding machine	U (21, 25)
Polishing machine	U (24, 28)
Riveting machine	U (14, 18)
Assembly	U (11, 12)

ulation model was very close to the real production level attained by the company. Our virtual production model produced 397.8 units of the product and real production counted with a production level of 400 units. Therefore, we can consider our simulation model valid and representative of the existing production system.

The company actual production environment can be described as chaotic. It has been impossible to meet customers' demand, and the stock of unfinished product, accumulated over time, between workstations is very high.

As previously mentioned our base model corresponds to the current production system, that includes a number of operations to be performed at specified workstations in a specified sequence. By running our simulation base model it will be possible to analyse the flow of the production line, identifying the process bottleneck, and determining the number of unfinished products through the production line accumulated per week. Our simulation was conducted considering a production demand of 5000 units, per week, that the company actually faces. To identify the number of unfinished products accumulated per week we have considered that, per day, the production line receives the necessary amount of raw material for a batch of 1000 units of our product. At the end of the first day of production we achieve an activity level of 398 finished products and 602 WIP items. On the second production day we add a new batch of raw material and continue to process the 602 WIP items, stocked between the line workstations, from the previous day. This procedure was then repeated along the following days of the week. This is, at the beginning of a new production day we receive a new batch of raw material and we continue the production flow of the WIP of the previous production day. The results of this simulation allowed us to analyse the production line behaviour during one week.

Gathered results showed that the production line bottleneck is the Cutting and Bending workstation (see Table 5), and that at the end of the week rather then producing the necessary production level of 5000 units, it was only possible to produce 4065 units of the product. The remaining 935 items are not finished and are located through the production line.

This initial analysis of the production system behaviour confirmed that the company is incurring production delays, since it is not possible to produce the necessary amount of units per week. This lead us to test new production scenarios, with different production strategies/procedures attempting to achieve production targets in a way that it is possible to face company's demand.

4.1 Scenarios with Intermediate Stock of WIP Items

Attempting to find a solution to increase the production level and meet the market demand, three different scenarios were analysed (see Fig. 3). In this set of simulation scenarios the use of intermediate stock of WIP items was considered. As the intermediate stocks will be stored after the bottleneck workstation, four stocking points were considered in the model, one for each of the remaining workstations on the production line.

The first stock position for WIP items was located after the Cutting and Bending machine and before the Sanding machine. Between the Sanding machine and the Polishing machine we will have a second point of intermediate stock, the third point of stock will be placed between the Polishing and the Riveting machines, and the last stocking point will be located after the Riveting machine and before the Assembly process.

These three scenarios will be simulated using the same computer model representation of the production system. The difference between them is the amount of stock (batch size) that is going to be considered in each of the scenarios. Table 2 illustrates the size of intermediate stock batch considered in each of the four stocking points for the analysed scenarios.

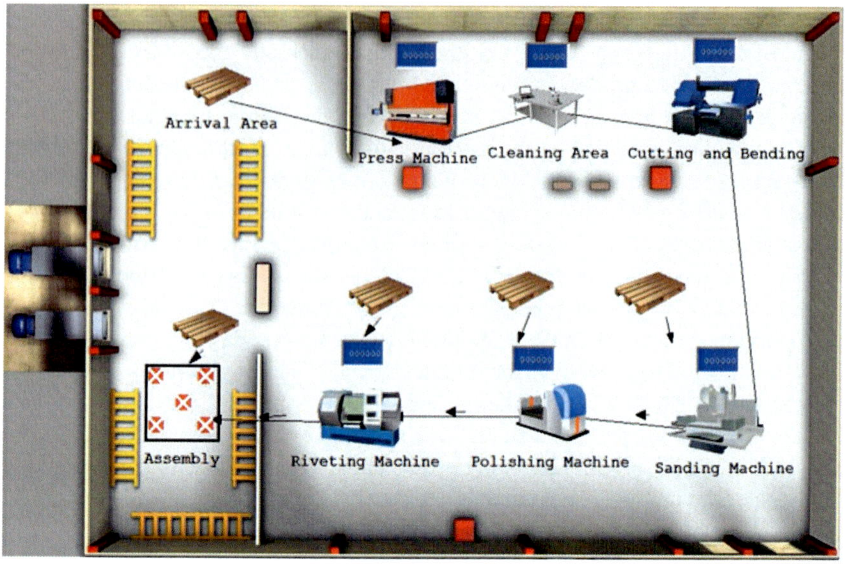

Fig. 3 Computer model for scenarios I, II, and III in ProModel® software environment

Table 2 Batch size of the intermediate stocks

	Batch size (units)		
	Scenario I	Scenario II	Scenario III
Stock position 1	75	50	80
Stock position 2	90	100	90
Stock position 3	200	200	200
Stock position 4	400	400	400

The aim of analysing these scenarios is to find the most suitable batch size of unfinished items, that are already in the production system, that together with the raw material that arrives every day, allows the production target of 1000 units, per day, needed to meet the actual demand of the company.

The results regarding our base model and these three scenarios will be later shown and discussed in Sect. 5.

The strategy of using WIP items in the production planning is dependent of the existence of those items. At the time of this research the stock level of unfinished products was very high, but in a future scenario without WIP items new solutions to increase production level must be conducted. In the next subsection, scenarios with the addition of new equipment will be simulated and analysed.

Table 3 Number of cutting and bending machines

	Number of machines
Base model	1
Scenario IV	2
Scenario V	3

4.2 Scenarios with the Addition of New Equipment

Aiming the elimination of the production line bottleneck and consequently increase the company's level of production, two scenarios will be considered. These scenarios may be considered as an alternative to be able to meet the weekly production target when the existent intermediate stocks cease to exist.

In Scenario IV we will double the actual number of cutting and bending machines (the production line bottleneck workstation), and in Scenario V we will triple the actual number of cutting and bending machines (see Table 3). Gathered results regarding these scenarios are shown and discussed in Sect. 5.

In the next subsection a new strategy is analysed. Taking into account the results of the previous scenarios we will consider a mixed strategy that includes the addition of new equipment together with the use of intermediate stock of WIP items.

4.3 Scenarios with a Mixed Strategy: Addition of New Equipment and with Intermediate Stock of WIP Items

This set of new scenarios is based on the hypothesis of acquisition a new cutting and bending machine associated with the use of intermediate stocks of WIP items. According to the results of the previous scenarios, that are shown and discuss in detail in Sect. 5, the adoption of a production planning based on this strategy only make sense after the company drain the production of the existent stock of the WIP items.

Because with the addition of a new cutting and bending machine it is still not possible to achieve the production target, with scenarios VI, VII and VIII we intend to find the smallest batch size of unfinished items that the company should produce, for example, in overtime production hours to meet the market's demand.

The stocking points described in Sect. 4.1 will also be considered in this set of scenarios. Table 4 shows the considered batch size of the WIP intermediate stocks in each of the four stock locations in each of the analysed scenarios.

Table 4 Batch size of the intermediate stocks

	Batch size (units)		
	Scenario VI	Scenario VII	Scenario VIII
Stock position 1	25	20	15
Stock position 2	50	20	15
Stock position 3	100	100	100
Stock position 4	200	200	200

5 Research Results and Discussion

In this section we present and discuss the results of our simulation model and all of the developed simulation scenarios. As already mentioned our simulation results involved the production of one single product, and considered a 8H45 min working period. We also consider an average of twenty three working days during a month, so 23 replications were programmed.

We will start to discuss the results of the base model (see Table 5) that describes the company's actual production system, and then the gathered results for all the implemented scenarios. Considering our base model, during the considered simulation time it was possible to produce 398 final products (pans). This activity production level may be considered very low since the daily batch size that it is necessary to produce should account for 1000 units. One may say that only 39.8% of the production level target is reached.

The simulation of the base model showed that the press machine spends 60.22% of the time blocked, and the cleaning area workstation also remains blocked 69.25% of the time. Besides that, it was possible to verify that the cutting and bending machine is operational during 99.82% of the time period. These results allowed us to confirm that the production line bottleneck is the workstation with the cutting and bending equipment. Consequently, subsequent workstations of the production line will be

Table 5 Simulation results for the base model

Workstation	Working time (%)	Blocked time (%)	Idle time (%)
Press machine	25.80	60.22	13.98
Cleaning area	30.63	69.25	–
Cutting and bending machine	99.82	–	0.18
Sanding machine	31.83	–	68.17
Polishing machine	35.94	–	64.06
Riveting machine	22.10	–	77.90
Assembly process	15.89	–	84.11

Table 6 Simulation results for Scenario III

Workstation	Working time (%)	Blocked time (%)	Idle time (%)
Press machine	15.05	76.78	8.17
Cleaning area	17.84	82.04	0.12
Cutting and bending machine	58.07	41.76	0.17
Sanding machine	24.87	35.81	39.33
Polishing machine	36.21	26.92	36.87
Riveting machine	33.36	15.91	50.73
Assembly process	39.96	–	60.04

idle during some periods, between 64.06% and 84.11% of the time, and intermediate stock level of WIP items will increase before the cutting and bending workstation.

The three first proposed scenarios were based on the base model and include the use of intermediate stocks of WIP items. Simulations results showed that with both scenarios I and II it was not possible to achieve the production target of 1000 units per day (5000 units per week). With scenario I we got a production level of 998 units, and with scenario II a production level of 988 units. A production level of exactly 1000 units was obtained using scenario III. Nevertheless, with the two first scenarios we were very close to reach the 1000 units batch target. Table 6 shows in detail the results for scenario III.

Considering scenario III the assembly process workstation is the only workstation that is not blocked for any period of time. The working time percentage values vary from 15.05% to 58.07% that corresponds to the press machine and the cutting and bending machine, respectively. Both the cleaning area and the cutting and bending machine were operational during almost all the time, but the idleness of the remaining workstations vary from a minimum of 8.17% to a maximum of 60.04%.

So, it was possible to face the external demand with the use of intermediate stocks, and without the addition of new equipment. The optimum batch size that the company should consider in each of the four stocking points is the one considered for Scenario III illustrated in Table 4.

Although the company has a huge amount of unfinished products, if we use the strategy modelled in scenario III to plan the production during a period of time, we will reduce or even eliminate the existent stock level of the WIP items. After some production time it is expected that the level of the WIP items cease or drastically decrease.

For this reason, two new scenarios were simulated. In these scenarios new equipment were added aiming to reduce or even eliminate the bottleneck of the production line and meet the existent demand.

Scenario IV includes the addition of a new cutting and bending equipment in the production line. In this scenario the blocked time of the press machine and the cleaning area was reduced to 20.75% and 36.99%, respectively. Both of the cutting

Table 7 Simulation results for scenario IV

Workstation	Working time (%)	Blocked time (%)	Idle time (%)
Press machine	51.39	20.75	27.85
Cleaning area	61.05	36.99	1.96
Cutting and bending machine	99.76	–	0.24
Cutting and bending machine II	99.63	–	0.37
Sanding machine	63.60	0.77	35.63
Polishing machine	71.77	–	28.23
Riveting machine	44.17	–	55.83
Assembly process	31.73	–	68.27

and bending machines were operational for almost one hundred percent of the time, but the idleness of the remaining equipment was reduced to 28.23% (minimum idleness) and up to 68.27% (maximum idleness). All the results concerning this scenario are shown in Table 7.

In scenario V, the addition of two cutting and bending machines was considered in the production line. Analysing the simulation results for this scenario it is possible to see that both the press machine and the cleaning area are not blocked anymore. The idleness of the remaining equipment has decreased even more to a minimum idleness of 9.28% and a maximum idleness of 59.93%. The remaining results for this scenario are shown in detail in the Table 8.

Concerning the production level, after a day of work, for both case scenarios, the following results were obtained: 795 units were produced using scenario IV, and 1004

Table 8 Simulation results for scenario V

Workstation	Working time (%)	Blocked time (%)	Idle time (%)
Press machine	64.55	–	35.45
Cleaning area	76.76	–	23.24
Cutting and bending machine	83.73	–	16.26
Cutting and bending machine II	83.72	–	16.27
Cutting and bending machine III	83.72	–	16.27
Sanding machine	80.23	1.37	18.40
Polishing machine	90.72	–	9.28
Riveting machine	55.76	–	44.24
Assembly process	40.07	–	59.93

Table 9 Simulation results for scenario VIII

Workstation	Working time (%)	Blocked time (%)	Idle time (%)
Press machine	43.40	33.08	23.52
Cleaning area	51.55	47.02	1.43
Cutting and bending machine	84.21	15.61	0.18
Cutting and bending machine II	84.06	15.65	0.29
Sanding machine	54.84	15.38	29.78
Polishing machine	63.26	13.40	23.34
Riveting machine	44.43	7.93	47.64
Assembly process	39.93	–	60.07

units were produced with scenario V production strategy. In summary only scenario V strategy allows the company to achieve the production target of 1000 units, so using the strategy of adding more machines the addition of one more equipment will not be sufficient and two more machines should be added.

Finally, three more scenarios were tested considering a mixed strategy that gathers the addition of new equipment and the use of intermediate stock of WIP items. At the end of a production day it was possible to produce 1012 units using scenario VI, 1002 units in scenario VII, and 1000 in scenario VIII. In what concerns the necessary amount of intermediate stock of the WIP items, for scenario VI it is necessary an early production level of 375 unfinished items, 340 units in scenario VII, and 330 units in scenario VIII.

Analysing the simulation results we may consider scenario VIII as our best option, since the production target of 1000 units is attained, and the level of the necessary intermediate stock of the WIP items is lower when compared with the necessary level for scenarios VI and VII. A detailed description of the simulation results in what concerns the percentage of the working time, the blocked time, and the idleness of each workstation is shown in Table 9.

6 Conclusions

This work analyses a production line of a metalworking company by means of the simulation technique. The goal was to find the bottleneck of a production line and to develop strategies to increase production level in order to face external demand.

The employment of simulation tools is becoming more and more significant when dealing with complex manufacturing environments because it is possible to model and compare the systems' behaviour in different scenarios, anticipating results, and consequently allowing the optimisation of the manufacturing systems without the associated risk of introducing significant changes in the real system.

The performed study identified the bottleneck on a production line, and highlighted the need to increase the production level to face the existing demand. Throughout this process, the analysis of our developed base model has revealed to be crucial.

Other eight different scenarios were also created: the first three scenarios included the possibility of using intermediate stocks of WIP items along the production line, two others showed production level improvements through the addition of new equipment, and the remaining scenarios included a mixed strategy with the use of intermediate stock of WIP items together with the addition of new equipment.

All the analysed scenarios confirmed the possibility of increase the daily production level. In particular, in the scenario III and scenario VIII the production level attained the batch size of 1000 units demanded. In summary it will be possible to face market demand in a first stage using scenario III strategy, but in the future it may be necessary to add one more cutting and bending machine. Nevertheless, with one more cutting and bending machine it is still necessary to use overtime production to create the necessary level of intermediate WIP items.

Regarding future work, it is possible to expand the developed model, integrating other products manufactured by the company in an attempt to improve other aspects of the production system. Also some new scenarios may be developed and analysed.

References

1. Banks, J.: Introduction to simulation. In: 2000 Winter Simulation Conference. IEEE (Cat. No. 00CH37165), pp. 9–16 (2000)
2. Carson, Y., Maria, A.: Simulation optimization: methods and applications. In: 1997 Winter Simulation Conference; IEEE **1**, 118–126 (1997)
3. Chiang, S.Y., Kuo, C.T., Meerkov, S.M.: c-Bottlenecks in serial production lines: identification and application. Math. Probl. Eng. **7**(6), 543–578 (2001)
4. Guimarães, A.M.C., Leal, J.E., Mendes, P.: Discrete-event simulation software selection for manufacturing based on the maturity model. Comput. Ind. **103**, 14–27 (2018)
5. Hassan, S.A., Abdelsalam, H.M.: An exploratory simulation model for contraception methods' supply chain in Egypt. Procedia Technol. **16**, 1403–1410 (2014)
6. Hernández, M.A.G., Librantz, A.F.H.: Simulación de eventos discretos de la cadena logística de exportación de commodities. Ingeniare. Revista chilena de ingeniería **22**, 257–262 (2014)
7. Iucksch, A.M.: Simulação de sistemas de gestão de produção em manufatura sazonal. Rio Grande do Sul, Brasil (2005)
8. La Mota, I.F., Guasch, A., Mota, M.M., Piera, M.A.: Integration of SIMIO with Coloured Petri Nets. Springer, Berlin (2017)
9. Law, A.M., Kelton, W.D., Kelton, W.D.: Simulation modeling and analysis, 2nd edn. McGraw-Hill, New York (1991)
10. Liu, C.M., Lin, C.L.: Performance evaluation of unbalanced serial production lines. Int. J. Prod. Res. **32**(12), 2897–2914 (1994)
11. Loaiza, M.E.B., Sarmiento, G.C., Correa, J.H.R.: Productividad en una celda de manufactura flexible simulada en promodel utilizando path networks type crane. Revista Tecnura **19**(44), 133–144 (2015)

12. Lu, M., Wong, L.-C.: Comparison of two simulation methodologies in modeling construction systems: Manufacturing-oriented PROMODEL vs. construction-oriented SDESA. Autom. Constr. **16**, 86–95 (2007)
13. Mourtzis, D., Doukas, M., Bernidaki, D.: Simulation in manufacturing: review and challenges. Procedia CIRP **25**, 213–229 (2014)
14. Müller, M.C., Dalberto, R., Rebelatto, T., Botassoli, G.T., Furtado, J.C.: Simulação computacional discreta em uma linha de produção de alimentos embutidos. Tecno-Lógica **19**, 49–56 (2015)
15. Sanjaya, S., Perdana, T.: Logistics system model development on supply chain management of tomato commodities for structured market. Procedia Manuf. **4**, 513–520 (2015)
16. Shannon, R.E.: Introduction to the art and science of simulation. In: Proceedings of the 1998 Winter Simulation Conference (Cat. No. 98CH36274), vol. 1, pp. 7–14 (1998)
17. Smith, J.S., Sturrock, D.T., Kelton, W.D.: Simio and simulation: modeling, analysis, applications. Simio LLC (2018)
18. White, K.P., Ingalls, R.G.: Introduction to simulation. In: Proceedings of the 2009 Winter Simulation Conference (WSC), vol. 1, pp. 12–23. IEEE (2009)

Design and Planning of Green Supply Chains with Risk Concerns

Cátia da Silva, Ana Carvalho, and Ana Paula Barbosa-Póvoa

Abstract Currently, legislative and social pressures led companies to widen their main goals and have been now considering not only economical but also environmental goals into their supply chain activities. In addition, the uncertainty associated with supply chains activities and the need to obtain more realistic values led to the need of addressing the associated financial risk. This paper explores this tendency and presents a mixed integer linear programming model for the design and planning of green supply chains that account for the economic and environmental concerns in the same objective function by monetizing environmental impacts. Also, the goal of minimizing the associated financial risk is considered. To deal with these two goals an augmented ε-constraint method is used to generate a pareto-optimal curve to determine the trade-off between the two objective functions. In this way, the present work contributes to the literature by providing a new model that considers the environmental impacts' monetization as well as models risk considering both economic and environmental performances of the supply chain. A case study is explored.

Keywords Supply chain · Risk management · Monetization · Uncertainty

1 Introduction

Supply chain (SC) management has been well-known for years in both academic and business communities. Taking into account that supply chain is a logistic system that covers a set of all activities from suppliers to customers, there is always high complexity associated. For that reason, it involves careful and efficient management in order to be possible to obtain satisfactory results for the company.

Considering the complex decisions, number of products, variables and entities involved, managing supply chain is a challenge, particularly if this management covers the design and planning of the supply chain. This management should be capable of optimizing customer value and achieving a competitive advantage in the market.

C. da Silva (✉) · A. Carvalho · A. P. Barbosa-Póvoa
CEG-IST, Instituto Superior Técnico, University of Lisbon, Av. Rovisco Pais, 1049-001 Lisbon, Portugal
e-mail: catia.silva@tecnico.ulisboa.pt

S. Relvas et al. (eds.), *Operational Research*, Springer Proceedings in Mathematics & Statistics 374, https://doi.org/10.1007/978-3-030-85476-8_12

In the past, when supply chains were optimized, decision-makers were only concerned with the economic performance of the supply chain. Taking into account the increasing companies' competitiveness and governmental pressures, SC has extended its goals to consider environmental and social concerns as well [4]. The management of those economic, environmental and social concerns led to the appearance of the sustainable supply chain (SSC) concept, which is a long well-recognized area in the World Commission on Environment and Development [1].

In addition to the sustainability and responsiveness that supply chain management must obey, risk management is also a reality that needs to be accounted for. Risk management has always been a debated subject in the academic community, because of the different points of view that management can encompass, as well as because of the adequacy of the use of risk measures. However, when it comes to design and planning of supply chain, it is consensual among the scientific community that risk management should be addressed, since there are many uncertainties associated that can only be taken into account when considering the risk [4]. Thus, the design and planning models for minimization of supply chain costs need to become more holistic. Actually, most of the real supply chain problems have several uncertain aspects associated, such as demand, and raw material prices [6]. In this way, it is necessary to develop risk management tools that efficiently address the supply chain uncertainty. In fact, there are several examples of the application of risk measures in supply chain design and/or planning in the literature [3, 10]. Nevertheless, this is not applied neither in the environmental nor in the social performance of the SC. But is important to highlight that risk can impact multiple sustainability dimensions, namely economic, environmental and social performances. Therefore, there is the need to account for those performances when analysing risk in supply chains, particularly in green and in sustainable supply chains.

This work addresses this need and presents a mixed integer linear programming model (MILP) that accounts for the economic and environmental concerns in the same objective function by monetizing environmental impacts and considering a risk measure. The goal is to maximize the difference between economic and environmental performances while minimizing the associated risk.

2 Methodologies for Monetization of Environmental Impacts

Taking into account the assessment of environmental impacts, there are several methods to assess them. According to the literature, the most used methodology is life cycle assessment (LCA), which is composed by four main phases, namely the goal and scope definition, inventory analysis, life cycle impact assessment (LCIA), and life cycle interpretation.

LCIA encompasses four steps: the classification of flows into impact categories, the quantification of the impacts through a characterization process, the normalization and the application of weighting factors. This may include some difficulties in choosing the best method to use. The LCIA methods allow us to quantify environmental impacts, however, the majority of them evaluate the environmental impacts in points. Few of the methods apply monetization to environmental impacts. Monetization is the process of translating environmental impacts into monetary units. According to the European Commission [7], EPS 2000 is a quite complete method of monetization that has the uncertainties fully specified when compared to other LCIA methods.

For this reason, in this work, monetization is performed by using EPS. The environmental impacts are quantified in a monetary unit (in euros) through EPS, which is mainly derived from future costs (raw material depletion), direct losses (production) and willingness-to-pay (health, biodiversity, and aesthetic values). Considering the willingness-to-pay, EPS method evaluates the environmental impacts based on academic knowledge from the Organization for Economic Cooperation and Development (OECD), which correspond to the value that the inhabitants would be willing to pay to avoid those environmental impacts.

3 Problem Description and Model Characterization

For this work, it is considered a generic supply chain, where it is intended to study its design. This SC considers raw materials flow from suppliers to factories, where final products are produced. Then, these final products move to warehouses or straight to markets. Also, at the markets, end-of-life products may be recovered and sent to warehouses or right to factories so as to be remanufactured.

Considering the possible set of locations of SC entities, products within the SC, technologies involved, and transportation modes between entities, the main purpose is to obtain the network structure of the SC, supply and purchase levels, transportation network, entities' capacities production, storage and remanufacturing levels, supply flow amounts, and product recovery levels, with the aim of maximizing profit, minimizing environmental impacts and financial risk.

The model used to solve this problem is a mixed integer linear programming (MILP) model that is based on the model presented in [9]. This base model is extended to consider simultaneously two objective functions through the augmented -constraint method [8]. This extension considers the maximization of the difference between economic and environmental performances, and the minimization of the financial risk, considering demand uncertainty.

Product's demand uncertainty is modelled through a scenario tree approach, since this is a method that allows the discretization of stochastic data over the time horizon and can be adjusted during the planning horizon. In this scenario tree, a node N characterizes a possible state and the arcs represent the evolutions it may have. Each node has a specific probability and a path from the root to a leaf node represents a scenario.

In the model developed, Eq. (1) represents the first objective function, which is the maximization of the difference between the expected net present value (eNPV) and expected environmental impact (eEnvImpact). The economic performance is assessed through the eNPV (considered in Eq. (2)), which is obtained through the sum of each node's probability multiplied by the discounted cash flows (CF_{Nt}) in each period t and for each node N at a given interest rate (ir). These CF_{N_t} are obtained from de net earnings (difference between incomes and costs). The costs include raw material costs, product recovery costs, operating costs, transportation costs, contracted costs with airline or freight, handling costs at the hub terminal, inventory costs, and labour costs. Furthermore, for the last time period, it is considered the salvage values of the SC (FCI_γ). The environmental performance is assessed by the eEnvImpact (considered in Equation (3)), LCA is performed on the transportation modes, in the production technology involved, and on entities installed in the SC boundaries, using EPS 2015 (an updated version of EPS 2000). The Life Cycle Inventory is retrieved from the Ecoinvent database (assessed through SimaPro 8.4.0 software). The LCA results are expressed in Environmental Load Units (ELU) and used as input data (ei) in Eq. (3), particularly in the environmental impact of transportation (first term), in the environmental impact of entity (second term), and in the environmental impact of technology (third term). Considering that environmental impacts are in a monetary unit, it is necessary to use a similar methodology to the discounted cash-flows that result in the present value of expected future environmental impacts at a given interest rate (ir).

The second objective function is represented by Eq. (4), where financial risk is minimized, which is modelled through the adoption of conditional value at risk (CVaR), where VaR is the minimum difference between NPV and EnvImpact for a given confidence level (α). CVaR evaluates the likelihood of obtaining the difference between eNPV and eEnvImpact lower than value-at-risk (VaR). CVaR apart from being recognised as a coherent risk measure has been used in supply chain studies where it has been proved to be quite useful in supporting decision makers decisions [2, 5].

$$\max(eNPV - eEnvImpact) \tag{1}$$

$$eNPV = \sum_N pb_N \times \left(\sum_t \frac{CF_{Nt}}{(1+ir)^t} - \sum_\gamma FCI_\gamma \right) \tag{2}$$

$$\begin{aligned} eEnvImpact = \sum_N pb_N \times \Bigg(& \sum_{a,m,i,j,t} \frac{ei_{ac} pw_m d_{ij} X_{maijtN}}{(1+ir)^t} + \sum_i ei_{ic} YC_i \\ & + \sum_{m,i,g,t} \frac{ei_{mgc} pw_m (P_{mgitN} + R_{mgitN})}{(1+ir)^t} \Bigg) \end{aligned} \tag{3}$$

$$\begin{aligned} \text{min Risk} = \max CVaR_\alpha = \max \left(VaR - \frac{1}{1-\alpha} \sum_N pb_N \times \delta_N \right); \\ \delta_N \geq VaR - (NPV_N - EnvImpact_N) \end{aligned} \tag{4}$$

The model also involves a set of constraints, not described here due to space restrictions, that model the main problem characteristics: entities locations, capacities definition (e.g., supply; stock; production and transportation); demand satisfaction; technology constraints modelling process suitability and associated performance; transportation constraints considering transportation modes suitability and availability.

4 Case Study

The developed model is applied to an appliance components' producer based in Verona, Italy. The company's suppliers are located in Verona. Presently, the company owns a factory and a warehouse with enough capacity to meet the demand of their existing clients. These clients are located in three main European markets, namely in Spain, Italy, and Germany. If there is a significant increase in demand, the supply chain is not able to respond adequately. Currently, there are some new potential clients that are interested in knowing if the company's decision-makers are going to study different SC forms of expansion.

Taking into account the clients' locations, Leeds and Hannover are possible locations for opening new factories. The installation of these new entities might lead to important changes in the transportation modes, namely for clients that are located in non-European countries. In addition, it is important to notice that warehouses' locations close to the markets should be taken into account. Currently, there are two types of products (fp1 and fp2) that are being produced. For this production, it is necessary to use two different types of technology. Moreover, it is important to highlight that end-of-life products can be recovered and remanufactured into final products. Additionally, there are three means of transport that can be used: truck, boat, and plane. There are two types of trucks (Truck1 and Truck2) that are different in capacity, variable costs, depreciation rate, and vehicle consumption. In this way, there is the possibility of transporting only by truck or through a combination of truck, boat and plane. Product's demand uncertainty is considered.

5 Results

Taking into account the SC reality, two cases are studied to understand the different decisions that can be made, regarding the SC design and planning. Case A considers a stochastic approach since products' demand is uncertain but it is not considered any risk measure; whereas case B contemplates uncertainty in products' demand and studies the trade-offs between the expected value of (eNPV-eEnvImpact) and the associated financial risk, assessed through CVaR. Case B is examined for the extreme values (the highest and lowest associated risk). The eNPV and the eEnvImpact obtained in both cases (A and B) are shown in Table 1. For determining the VAR, a confidence level α of 95% is considered, based on a decision-maker decision.

Table 1 Results for the economic (eNPV) and environmental (eEnvImpact) performances

	Case A Risk-neutral	Case B High-risk	Case B Low-risk
eNPV (x10^4 €)	171026	171054	170931
eEnvImpact (x10^4 €)	128571	128588	129965
eNPV-eEnvImpact €	42455	42466	40966

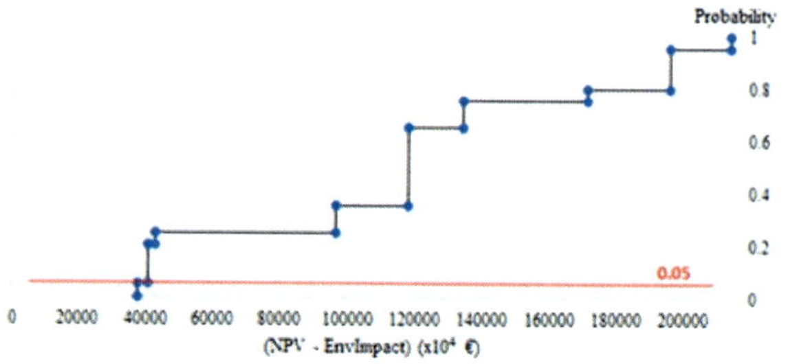

NPV - EnvImpact (x10^4 €)	Probability
38021	0.0500
40741	0.2000
42785	0.2500
96774	0.3500
118088	0.6500
134077	0.7500
171402	0.8000
195792	0.9500
214086	1.0000

Fig. 1 Cumulative distribution function to determine VaR

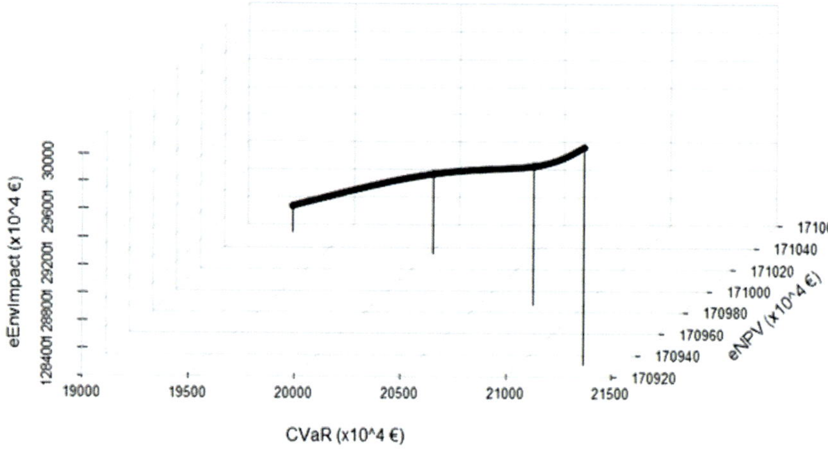

Fig. 2 Pareto curve using CVaR risk measure

For case B, assuming a confidence level of 95%, it is necessary to compute all the scenarios and assess the cumulative distribution function (Fig. 1). Thus, VaR is equal to 40741 $\times$ 10^4 €, which means that, at the end of the time horizon, the difference between eNPV and eEnvImpact for the design and planning of this SC is going to be, at least, this value. Accordingly, it is possible to obtain the Pareto curve assessed through the CVaR risk measure (Fig. 2).

When analysing the supply chain structure decisions, it can be seen in Table 2 that there are significant changes comparing cases A and B. In addition, besides

Table 2 Summary of the SC decisions considering the different cases performances

		Case A Risk-neutral	Case B High-risk	Case B Low-risk
Factories		Verona and Hannover	Verona and Hannover	Verona and Hannover
Warehouses		Verona and Sofia	Verona and Sofia	Verona and Sofia
Suppliers allocation		All factories are supplied by the closest supplier plus Leeds supplier	All factories are supplied by the closest supplier plus Leeds supplier	All factories are supplied by all suppliers
Production		Most production of fp1 is in the existing factory	Most production of fp1 is in the existing factory	Most production of fp1 is in the existing factory
		Most production of fp2 is divided between the two factories	Most production of fp2 is divided between the two factories	Most production of fp2 is divided between the two factories
Remanufacture		All remanufacture is in the existing factory	All remanufacture is in the existing factory	All remanufacture is in the existing factory
Inventory	fp1	45% in Verona and 55% in Sofia	45% in Verona and 55% in Sofia	fp1 and fp2 divided between Verona and Sofia
	fp2	60% in Verona and 40% in Sofia	60% in Verona and 40% in Sofia	fp1 and fp2 divided between Verona and Sofia
Transportation	Road	2 Truck1 and 14 Truck2	3 Truck1 and 14 Truck2	3 Truck1 and 15 Truck2
	Air	Not used	Not used	Not used
	Sea	Used	Used	Used

the already existent factory in Verona, all cases are going to have the same new factory installed (Hannover). Moreover, the results also show that in all cases there is the need to expand the existing capacity by opening a new warehouse. Concerning suppliers' allocation, it can be seen that in risk-neutral and high-risk cases the results show that all factories are supplied by the closest suppliers plus Leeds' supplier, which can be explained by the fact that the closest suppliers reach its capacity and it is necessary to use more distant ones. In contrast, in the low-risk case, where environmental impacts are higher, all factories are supplied by all suppliers, due to the balance between the lower costs of transportation and low costs of raw materials. Regarding intermodal transportation, and for all cases, sea option is preferred to air option, since sea transportation has a lower environmental impact. Considering a careful analysis into the results, it is important to highlight that network structure of the risk-neutral case is similar to the high-risk case, due to the fact that both SC entities and modes of transportation are similar, with some differences in capacities involved and in the product flows between entities.

6 Conclusion

This work presents a mixed integer linear programming model for the design and planning of SC that accounts for the economic and environmental concerns in the same objective function by monetizing environmental impacts and minimizing the associated risk under uncertainty on the products demand.

For the risk assessment, conditional value-at-risk was considered, particularly because it is the most common risk measure used in the literature and there is a consensus on their application in this area. To show the trade-offs between (eNPV - eEnvImpact) and risk, a Pareto optimal curve was developed through the application of the augmented ε-constraint method.

It is important to highlight that this analysis was only possible, due to the process of monetization, which allow us to quantify the environmental impacts in a monetary unit, namely in euros. In addition, monetization enables us to include in the same objective function both economic and environmental performances. Moreover, from the analysis done, it is clear that results are influenced by differences in risk strategies and this proves the importance of risk management in solving real-life problems.

For future work, there is the need to develop this topic, particularly to evolve monetization approaches so as to be an effective and reliable alternative to evaluate environmental impacts and better understand how to relate environmental costs with other supply chain costs. Additionally, an extension of this work should consider other risk measures in order to better conclude on its adequacy.

Acknowledgements This work was supported by Portugal 2020, POCI-01-0145-FEDER-016418 by UE/FEDER through the program COMPETE2020 and FCT and Portugal 2020 under the project PTDC/EGE-OGE/28071/2017 and Lisboa - 01-0145-FEDER-28071.

References

1. Report of the World Commission on Environment and Development: Our Common Future (1987). https://www.are.admin.ch/dam/are/en/dokumente/nachhaltige_entwicklung/dokumente/bericht/our_common_futurebrundtlandreport1987.pdf.download.pdf/our_common_futurebrundtlandreport1987.pdf
2. Amorim, P., Alem, D., Almada-Lobo, B.: Risk management in production planning of perishable goods. Ind. Eng. Chem. Res. **52**(49), 17,538–17,553 (2013). https://doi.org/10.1021/ie402514c
3. Baptista, S., Barbosa-Póvoa, A.P., Escudero, L.F., Gomes, M.I., Pizarro, C.: On risk management of a two-stage stochastic mixed 0–1 model for the closed-loop supply chain design problem. Eur. J. Oper. Res. **274**(1), 91–107 (2019). https://doi.org/10.1016/j.ejor.2018.09.04

4. Barbosa-Póvoa, A.P., da Silva, C., Carvalho, A.: Opportunities and challenges in sustainable supply chain: An operations research perspective. Eur. J. Oper. Res. **268**(2), 399–431 (2018). https://doi.org/10.1016/j.ejor.2017.10.03
5. Carneiro, M.C., Ribas, G.P., Hamacher, S.: Risk management in the oil supply chain: A cvar approach. Ind. Eng. Chem. Res. **49**(7), 3286–3294 (2010). https://doi.org/10.1021/ie901265n
6. Cristobal, M.P., Escudero, L.F., Monge, J.F.: On stochastic dynamic programming for solving large-scale planning problems under uncertainty. Comput. Oper. Res. **36**(8), 2418–2428 (2009). https://doi.org/10.1016/j.cor.2008.09.009
7. Commission, European: Joint Research Centre- Joint Research Centre - Institute for Environment and Sustainability: International Reference Life Cycle Data System (ILCD) Handbook- Recommendations for Life Cycle Impact Assessment in the European context. Publications Office of the European Union (2011). https://doi.org/10.278/33030
8. Mavrotas, G.: Effective implementation of the ϵ-constraint method in multi-objective mathematical programming problems. Appl. Math. Comput. **213**(2), 455–465 (2009). https://doi.org/10.1016/j.amc.2009.03.037
9. da Silva, C., Barbosa-Póvoa, A.P., Carvalho, A.: Sustainable supply chain: monetization of environmental impacts, pp 773–778 (2018). https://doi.org/10.1016/B978-0-444-64235-6.50136-4
10. Zeballos, L.J., Gomes, M.I., Barbosa-Povoa, A.P., Novais, A.Q.: Addressing the uncertain quality and quantity of returns in closed-loop supply chains. Comput. Chem. Eng. **47**, 237–247 (2012). https://doi.org/10.1016/j.compchemeng.2012.06.034

Benchmarking Smart Grid Research & Development Engagement by European Distribution System Operators

Micael Simões, Rogério Rocha, and Ana Camanho

Abstract Technological developments related to renewable energy led to a decrease on the prices of generation and allowed the penetration of distributed energy resources in power systems. This context, combined with other factors, such as the development of electric vehicles, enabled the rapid evolution of Smart Grids. As a consequence, Distribution System Operators (DSOs) have been investing in this field to keep up with its deployment. This work presents a case study that compares a set of European DSOs regarding their investment in Smart Grid projects. The methodology underlying this study is based on the construction of composite indicators using the Data Envelopment Analysis technique. Furthermore, we evaluate the evolution in the DSOs performance between 2013 and 2017 using a Malmquist index. The results are discussed in the light of their contribution to the definition of public policies in the energy field.

Keywords Smart grids · Distribution system operators · Data envelopment analysis · Composite indicators · Distributed generation · Malmquist index

1 Introduction

Electrical power systems are on a fast up-to-bottom change worldwide. Some of the factors responsible for these changes include the high penetration of the Distributed Energy Resources (DER), the rising share of generation from Renewable Energy Sources (RES) and the rapid growth in the number of Electric Vehicles (EV).

M. Simões · R. Rocha
Center for Power and Energy Systems, INESC-TEC, Faculdade de Engenharia, Universidade do Porto, Porto, Portugal
e-mail: up201809130@fe.up.pt

R. Rocha
e-mail: up200804930@fe.up.pt

A. Camanho (✉)
Faculdade de Engenharia, Universidade do Porto, Porto, Portugal
e-mail: acamanho@fe.up.pt

S. Relvas et al. (eds.), *Operational Research*, Springer Proceedings in Mathematics & Statistics 374, https://doi.org/10.1007/978-3-030-85476-8_13

Consequently, the changes required to the electricity infrastructure led to the development of Smart Grids (SGs), as a means to accommodate to the new requirements securely and economically, without sacrificing the quality of service provided to the end-consumer [1].

SGs can be described as an upgraded electricity network, that allows a two-way communication and power exchange between electricity suppliers and consumers. In the last few years, initiatives in the SGs field have been growing, with different objectives and application scopes. Substantial investments have been made in Research and Development (R&D), demonstration and deployment of SG projects. In this context, the assessment of DSOs' engagement in R&D is crucial to evaluate their efforts for the achievement of the envisioned SGs targets [2]. A benchmark study at this stage can be helpful for the design and refinement of DSOs' research strategies. The findings of this exploratory study can highlight the differences in relative performance among the DSOs, such that operators with lower levels of R&D investment may infer their readiness for the SG era and develop efforts to converge towards best practice levels.

In this research, we apply Data Envelopment Analyis (DEA) to a set of European DSOs that have been actively developing research, development and innovation work in the SGs field. The main aim of this benchmark study is to monitor past achievements and assess the DSOs' research strategies for upcoming challenges related to SGs. Given the multitude of SG areas, and the different investment strategies followed by the DSOs under analysis, traditional evaluation methodologies are not suitable to conduct a relative assessment of the DSOs' performance. For this purpose, we propose the construction of a composite indicator using a methodology organized in three stages. In the first stage we benchmark DSOs in five different categories related to the purposes and areas of application of smart grids. This is followed by the construction of an aggregate measure of performance, providing an overview of the relative position of the DSOs, considering all dimensions. Finally, we explore the trends in the evolution of performance in recent years (between 2002 and 2017).

This study is based on data made available by European Commission's Joint Research Center (JRC), through their Smart Grid Projects Inventory database [3]. We expect to contribute to the convergence in DSOs performance in the European space, motivating continuous improvement to close the gap between the leaders and the other DSOs that started the quest for excellence in SGs at a latter stage, or that have eventually followed less successful paths.

2 Methodology

2.1 Construction of Composite Indicators Using DEA

Data Envelopment Analysis (DEA), first introduced by Charnes, Cooper & Rhodes [4], is nowadays the most widely used method for comparative efficiency assessments. DEA is a technique used for the evaluation of performance of Decision Making Units (DMUs), in the presence of multiple indicators representing inputs and outputs of a production process [5].

The application of DEA for performance assessments has several advantages relatively to other methods. DEA is a non-parametric technique that does not require the knowledge of the functional form of the efficient frontier (or production function). It is also able to handle multiple inputs and multiple outputs measured in different scales. It provides a single summary measure of efficiency for each DMU, relatively to other peer DMUs under analysis [6].

Instead of the traditional DEA paradigm considering the transformation of inputs (resources) into outputs, in this paper we consider the paradigm underlying the construction of composite indicators (CI) for the evaluation of DSOs performance. Composite Indicators (CIs) constitute an aggregation of several indicators into a summary measurement of performance [7].

CIs can be estimated using DEA models, under the Benefit-of-the-Doubt (BoD) approach, a term coined by Cherchye et al. [8]. It uses a single dummy input, equal for all the DMUs, and multiple outputs that represent individual performance indicators to be combined in the composite measurement. Following Koopmans [9], Cook and Kress [10], Lovell [11] and Lovell et al. [12], we can have a unitary input underlying the evaluation of every DMU, interpreted as a "helmsman" attempting to steer the operators towards the maximization of outputs (i.e., in our case the outputs represent the DSOs' performance in the smart grids research field).

By considering a unitary input level for all DMUs in the original DEA model [4] and an output orientation, we obtain the model presented in (1), proposed in [13], which can be used to obtain a CI score according to the "benefit-of-the-doubt" paradigm. This model endogenously selects the weights that maximize the CI score for the DMU under assessment. Thus, each unit is always shown in its best possible light.

$$
\begin{aligned}
\frac{1}{\text{CI}} &= \text{Min } v \\
\text{s.t.} \sum_{r=1}^{s} & u_r y_{rj_0} = 1 \\
\sum_{r=1}^{s} & u_r y_{rj} - v \leq 0, \quad j = 1, \ldots, n \\
u_r &\geq \epsilon, \quad r = 1, \ldots, s \\
v &\geq 0
\end{aligned}
\tag{1}
$$

In model (1), j_0 corresponds to the DMU under evaluation from the set $j = 1, \ldots n$. y_{rj} is the r_{th} output ($r = 1, \ldots, s$) of the j_{th} DMU. u_r is the pure weight given to the r_{th} output and v the pure weight given to the dummy input. These weights are the decision variables of the linear programming model. The mathematical infinitesimal

ϵ is used to ensure that all outputs included in the model are taken into account in the performance evaluation. The Composite Indicator (CI) score is the inverse of the value of the weight v assigned to the dummy input, i.e., $\text{CI} = \frac{1}{v}$. The value of the CI is equivalent to the relative efficiency score associated with the objective function of a DEA formulation.

The formulation shown in (1) has an output orientation, which is more consistent with the goal of the performance assessment than the input oriented BoD model of [8], enabling an easier implementation of weight restrictions. It also allows a direct estimation of the targets on the frontier of the production possibility set.

Flexible weighting systems, adopting the BoD paradigm, include the possibility of specifying weights for each indicator that are different across individual operators. The flexibility in choosing the weights assigned to each indicator allows maximising the performance score of the DSO under assessment. According to Cherchye et al. [8], more (or less) weight should be attributed to those indicators in which the operator performed better (or worse), assuming that better performance results were achieved on the indicators that are more important for the operator under assessment. This means that the weights that are applied to the data are actually based on observed performance data, estimated endogenously using optimization techniques. However, although this weight flexibility can be seen as a strength of BoD evaluation, it can also be a weakness, as the operators may assign judicious weights that may in practice even ignore a few outputs in the performance assessment.

An important procedure to enhance discrimination in the performance evaluation, avoiding the occurrence of zero weights and reflecting the preference structure among the variables chosen for the evaluation of the DSOs, involves the use of weight restrictions. In our empirical analysis, we imposed restrictions to the indicators defined as shown in (2) and (3). These restrictions are imposed to virtual weights, as originally proposed by Wong and Beasley [14], but work as Assurance Regions type I (Thompson et al. [15]). This requires resorting to an artificial observation equal to the sample mean for each indicator considered, as proposed by Zanella et al. [13].

$$\frac{u_r \overline{y}_r}{\sum_{r=1}^{s} u_r \overline{y}_r} \geq \alpha_r \tag{2}$$

$$\sum_{r \in S_k} u_r \overline{y}_r \geq \sum_{r \in S_z} u_r \overline{y}_r, \text{ for } S_k \neq S_z \tag{3}$$

In expressions (2) and (3), $\overline{y}_r$ is a parameter equal to the sample mean for a given output indicator r for which the decision maker wishes to specify a restriction to its weight. u_r is the pure weight assigned to output r. α_r is a parameter, expressed in percentage, defining the minimum bound for the importance attached to output r.

In practice, whilst expression (2) imposes a minimum value for the relative importance of a given output in the overall performance evaluation, expression (3) imposes an order between the relative value of two subsets of outputs (S_k and S_z) chosen by a decision maker. The formulation shown in (3) states that the total virtual weight, expressed in percentage terms, assigned to the indicators from subset S_k must be greater than the total virtual weight assigned to the indicators from subset S_z.

One advantage of using these types of restrictions is that they allow for a fair comparison among the different outputs, even for indicators measured in different scales.

2.2 Framework for the Performance Evaluation

The European Commission's JRC categorizes Smart Grid projects in four main areas, according to their main purposes and areas of application. These are:

- SNM: Projects focused on increasing the operational flexibility of the electricity grid through enhanced grid monitoring and control capabilities, e.g. projects related with the installation of network monitoring and control equipment and fast and real-time data communications.
- DSM: Includes two main project types: demand-response (projects that aim at shifting electricity consumption to another point in time) and energy efficiency (projects that aim at decreasing the level of electricity consumption, while providing the same energy service, maintaining comfort levels).
- DG&S: Projects in this area are focused on advanced control schemes and information and communications solutions for integrating distributed generation and energy storage into the distribution network, while ensuring system reliability and security.
- e-Mobility: Projects focused on the smart integration of electric vehicles and plug-in hybrid electric vehicles into the distribution network.

In addition to these four main areas, JRC also considers an additional category, denominated "Others", where all of the remaining project types are included. Each project in JRC database is also categorized according to its regional scope (multinational or national), project stage (demonstration or R&D) and the role of each DSO in the team (leader or partner). These are important aspects to take into account, as multinational and demonstration projects are considered to have a greater contribution to SG development than national and R&D projects. In a similar manner, projects leadered by the DSO are also considered to be more relevant than projects where the DSO assumes a partner role.

The framework proposed for conducting the performance evaluation is composed of three stages. In the first stage, we compute for all DSOs a composite indicator for each SG area, representing the DSO's performance in that area. In the second stage, a single summary indicator is obtained to reflect the DSO's overall performance in the SGs field. Finally, a Malmquist Index is computed to assess the evolution of the DSOs performance during the period 2002–2017.

2.2.1 First-Stage—DSOs' Performance per Smart Grid Area

In the first stage of the framework, we obtain a measure of performance per smart grid area considering all projects in which they were engaged between 2002 and 2017. Therefore, the results of the first stage consist of a set of five indicators, one for each SG area considered.

When a large number of indicators are used in the assessment, considering only a relatively small number of DMUs, several DMUs can appear as efficient due to the lack of discrimination power in the model specified. In order to reduce the number of variables involved in the evaluation of the DSOs, a total of 6 indicators were carefully chosen for the first-stage. Regarding the number of projects, the model included four indicators: number of multinational demonstration projects as leader (n_{MDL}) and partner (n_{MDP}), number of multinational R&D projects (n_{MR}) and number of national projects (n_{N}). Regarding the investments, two variables were chosen: investment in multinational (inv_{M}) and national projects (inv_{N}). This selection of output variables was conducted in such a manner that all the data available for each DSO was used, without information overlap in the specified variables.

2.2.2 Second-Stage—DSOs' Performance on Smart Grid R&D Engagement

The result of the second stage of the framework is a summary measure of overall performance describing the DSO engagement in SG R&D projects. This CI measure is also obtained using the DEA technique. The input of the second stage consists of the results of the first-stage, i.e., the performance in each of the five SG areas considered for each DSO.

2.2.3 Third-Stage—Evolution of Performance over Time

To evaluate the DSO's change in performance over time, a Malmquist Index (MI) was computed, according to the formulation proposed by Fare et al. [16]. This approach allows the decomposition of productivity change into two components: Efficiency Change (EC) and Frontier Shift (FS).

Two time periods were analyzed, 2002–2013 and 2002–2017, with the objective of evaluating the state of the R&D work at the end of each time period. For this purpose, we considered the cumulative investment in projects until the end of each period analyzed.

3 Case Study Presentation

The DSOs were selected taking into account the availability of data in the JRC database and the country of origin. All DSOs belong to European Union countries, and in the period analysed were engaged in projects for at least 2 of the 5 areas considered. This resulted in the selection of 23 DSOs. Figure 1 identifies the DSOs included in the sample, and illustrates the investments and number of projects undertaken for the 5 smart grid areas until the year 2017.

Figure 1 shows that Iberdrola, ERDF and RWE are the most active DSOs in terms of SG developments, both in terms of investments and number of projects conducted. It is also possible to see that for most DSOs, the SNM, DSM and DG&S categories were the preferred investment areas.

In Tables 1 and 2 is listed the input data for the first stage of the analysis.

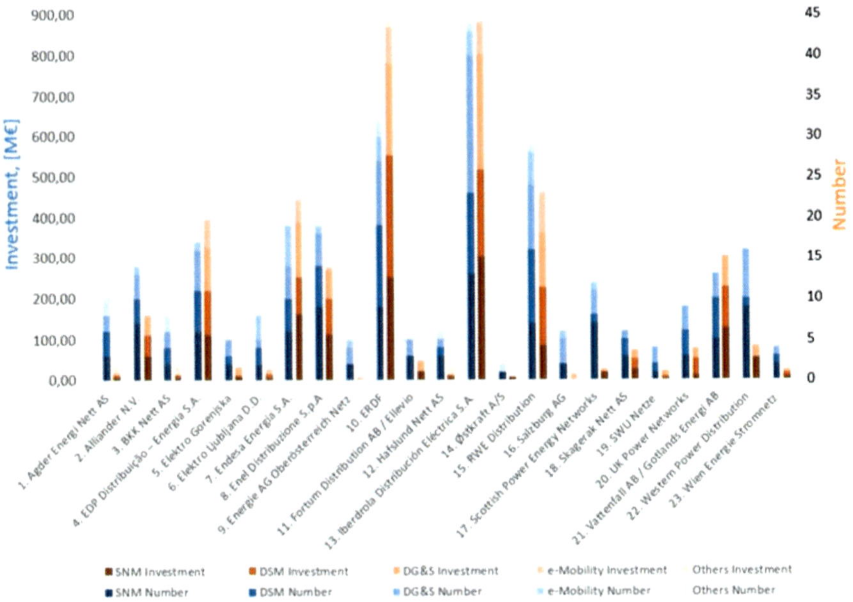

Fig. 1 Total investment and number of projects of the DSOs in each smart grid area

Table 1 First-stage data. Number of projects conducted by the DSOs in the period 2002–2017, disaggregated per area

DSO	SNM				DSM				DG&S				e-Mobility				Others			
	n_{MDL}	n_{MDP}	n_{MR}	n_N	n_{MDL}	n_{MDP}	n_{MR}	n_N	n_{MDL}	n_{MDP}	n_{MR}	n_N	n_{MDL}	n_{MDP}	n_{MR}	n_N	n_{MDL}	n_{MDP}	n_{MR}	n_N
1	0	1	1	1	0	1	2	0	0	0	1	1	0	0	0	0	0	0	1	1
2	0	2	4	1	0	1	1	1	0	2	0	1	0	1	0	0	0	0	0	0
3	0	1	1	0	0	1	1	0	0	0	1	1	0	0	0	0	0	0	1	1
4	0	3	2	1	0	3	1	1	0	3	1	1	0	0	0	1	0	0	0	0
5	0	0	2	0	0	1	0	0	0	2	0	0	0	0	0	0	0	0	0	0
6	0	1	1	0	0	1	1	0	0	0	1	0	0	2	1	0	0	0	0	0
7	0	2	2	2	0	2	1	1	0	1	0	3	0	1	2	2	0	0	0	0
8	1	2	4	2	1	2	1	1	0	2	1	1	0	0	0	1	0	0	0	0
9	0	0	0	2	0	0	0	0	0	0	0	2	0	0	0	1	0	0	0	0
10	2	1	1	5	2	2	1	5	2	0	1	5	0	1	0	2	0	1	1	0
11	0	1	1	1	0	0	0	0	0	1	0	1	0	0	0	0	0	0	0	0
12	0	1	0	2	0	1	0	0	0	0	0	1	0	0	0	0	0	0	0	1
13	3	3	4	3	2	3	3	2	2	3	9	3	0	1	1	1	0	0	1	0
14	0	0	1	0	0	0	0	0	0	0	0	0	0	0	0	0	0	0	0	1
15	2	1	3	1	3	1	4	1	2	1	3	2	0	1	3	0	0	0	1	0
16	0	0	0	2	0	0	0	0	0	0	1	2	0	0	0	1	0	0	0	0
17	0	0	0	7	0	0	0	1	0	0	0	3	0	0	0	1	0	0	0	0
18	0	1	1	1	0	1	1	0	0	0	1	0	0	0	0	0	0	0	0	0
19	0	0	1	0	0	0	1	0	0	0	1	1	0	0	0	0	0	0	0	0
20	0	0	2	1	0	1	0	2	0	0	1	2	0	0	0	0	0	0	0	0
21	1	4	0	0	0	5	0	0	0	3	0	0	0	0	0	0	0	0	0	0
22	0	0	1	8	0	0	0	1	0	0	0	6	0	0	0	0	0	0	0	0
23	0	0	0	2	0	0	0	1	0	0	0	1	0	0	0	0	0	0	0	0

Table 2 First-stage data. Investment in projects conducted by the DSOs in the period 2002–2017, disaggregated per area (in million €)

DSO	SNM		DSM		DG&S		e-Mobility		Others	
	inv_M	inv_N	inv_M	inv_N	inv_M	inv_N	inv_M	inv_N	inv_M	inv_N
1	6.35	0.43	6.90	0.00	2.12	0.88	0.00	0.00	1.27	17.50
2	59.66	1.00	48.02	1.00	46.06	1.00	5.11	0.00	0.00	0.00
3	6.39	0.00	6.39	0.00	2.16	0.88	0.00	0.00	1.27	17.50
4	42.18	70.00	36.40	70.00	37.04	70.00	0.00	70.00	0.00	0.00
5	10.83	0.00	2.84	0.00	18.17	0.00	0.00	0.00	0.00	0.00
6	7.22	0.00	7.22	0.00	2.27	0.00	11.61	0.00	0.00	0.00
7	63.56	101.00	26.71	61.00	34.85	101.36	55.31	0.86	0.00	0.00
8	91.80	21.71	79.40	7.40	62.34	7.40	0.00	7.40	0.00	0.00
9	0.00	2.50	0.00	0.00	0.00	2.50	0.00	1.25	0.00	0.00
10	128.52	122.50	176.66	125.70	89.09	136.20	71.00	19.00	15.29	0.00
11	3.89	19.88	0.00	0.00	3.40	19.88	0.00	0.00	0.00	0.00
12	4.24	3.91	4.24	0.00	0.00	0.88	0.00	0.00	0.00	17.50
13	167.28	134.00	122.65	94.00	149.81	134.00	46.74	34.00	2.40	0.00
14	8.01	0.00	0.00	0.00	0.00	0.00	0.00	0.00	0.00	1.32
15	82.65	4.13	108.59	31.85	97.38	35.98	99.17	0.00	7.84	0.00
16	0.00	2.50	0.00	0.00	6.62	2.50	0.00	1.25	0.00	0.00
17	0.00	19.19	0.00	0.06	0.00	4.80	0.00	0.06	0.00	0.00
18	25.24	2.01	25.24	0.00	21.00	0.00	0.00	0.00	0.00	0.00
19	4.59	0.00	4.59	0.00	4.59	8.70	0.00	0.00	0.00	0.00
20	10.38	2.10	3.80	36.10	5.62	18.95	0.00	0.00	0.00	0.00
21	127.57	0.00	98.37	0.00	75.51	0.00	0.00	0.00	0.00	0.00
22	6.56	48.39	0.00	0.34	0.00	26.62	0.00	0.00	0.00	0.00
23	0.00	9.68	0.00	8.00	0.00	8.00	0.00	0.00	0.00	0.00

4 Discussion of Results

4.1 First-Stage Results—DSOs' Performance per SG Area

The analysis of performance per SG area was based on the use of model (1), with two sets of weight restrictions. The first group of restrictions established a minimum relative weight for each of the five areas specified. These constraints, formulated according to expression (2), were implemented with the bounds shown in (4). These bounds were selected after discussion with experts in the field.

$$
\begin{aligned}
u_{n_{\text{MDL}}}\overline{y}_{n_{\text{MDL}}} &\geq 0.15\sum_{r=1}^{s}u_r\overline{y}_r \\
u_{n_{\text{MDP}}}\overline{y}_{n_{\text{MDP}}} &\geq 0.1\sum_{r=1}^{s}u_r\overline{y}_r \\
u_{n_{\text{MR}}}\overline{y}_{n_{\text{MR}}} &\geq 0.05\sum_{r=1}^{s}u_r\overline{y}_r \\
u_{n_{\text{MN}}}\overline{y}_{n_{\text{MN}}} &\geq 0.025\sum_{r=1}^{s}u_r\overline{y}_r \\
u_{inv_{\text{M}}}\overline{y}_{inv_{\text{M}}} &\geq 0.125\sum_{r=1}^{s}u_r\overline{y}_r \\
u_{inv_{\text{N}}}\overline{y}_{inv_{\text{N}}} &\geq 0.05\sum_{r=1}^{s}u_r\overline{y}_r
\end{aligned}
\tag{4}
$$

The second group of constraints intends to reflect the generalized preference for international projects over national projects, projects in a demonstration stage over R&D, and projects with a leading role over partnership. These weight restrictions were implemented as shown in (5).

$$
\begin{aligned}
& u_{n_{\text{MDL}}}\overline{y}_{n_{\text{MDL}}} + u_{n_{\text{MDP}}}\overline{y}_{n_{\text{MDP}}} \geq u_{n_{\text{MR}}}\overline{y}_{n_{\text{MR}}} \\
& u_{n_{\text{MDL}}}\overline{y}_{n_{\text{MDL}}} \geq u_{n_{\text{MDP}}}\overline{y}_{n_{\text{MDP}}} \\
& u_{n_{\text{MDL}}}\overline{y}_{n_{\text{MDL}}} + u_{n_{\text{MDP}}}\overline{y}_{n_{\text{MDP}}} + u_{n_{\text{MR}}}\overline{y}_{n_{\text{MR}}} \geq u_{n_{\text{N}}}\overline{y}_{n_{\text{N}}} \\
& u_{inv_{\text{M}}}\overline{y}_{inv_{\text{M}}} \geq u_{inv_{\text{N}}}\overline{y}_{inv_{\text{N}}}
\end{aligned}
\tag{5}
$$

The results obtained for the year of 2017 are shown in Table 3.

From the analysis of the table, it is possible to see that Iberdrola Distribución, ERDF and RWE Distribution are the DSOs with best relative performance in most of the areas. This result could be expected, given that these DSOs had the highest number of projects and the largest investments in the SGs field.

Iberdrola is the DSO with best relative performance for the the SNM and DG&S areas. ERDF and RWE are the benchmarks in the DSM area, closely followed by Iberdrola with 95.97%. In the e-Mobility area, the DSOs with best-practices are Elektro Ljubljana, Endesa Energía, ERDF and RWE. In the "Others" area, Agder Energi Nett, BKK Nett and ERDF are the best-performing DSOs. This fact highlights the need for obtaining an overall measure of DSOs performance, considering the whole spectrum of SG areas, as different DSOs prioritize different areas within the SG field.

Table 3 First-stage data. Investment in projects conducted by the selected DSOs until the year 2017, disaggregated per area

DSO	CI_{SNM}(%)	CI_{DSM}(%)	$CI_{DG\&S}$(%)	$CI_{e-Mobility}$(%)	CI_{Others}(%)
1. Agder Energi Nett AS	10.44	14.05	7.26	0.00	100.00
2. Alliander N.V.	29.78	21.33	21.56	50.00	0.00
3. BKK Nett AS	7.86	9.34	7.26	0.00	100.00
4. EDP Distribuição Energia S.A.	29.29	31.67	29.98	50.00	0.00
5. Elektro Gorenjska	8.60	5.35	12.24	0.00	0.00
6. Elektro Ljubljana D.D.	7.92	9.41	4.09	100.00	0.00
7. Endesa Energía S.A.	37.07	25.31	32.85	100.00	0.00
8. Enel Distribuzione S.p.A	50.91	49.45	28.61	50.00	0.00
9. Energie AG Oberösterreich Netz	7.58	0.00	8.20	50.00	0.00
10. ERDF	77.47	100.00	87.82	100.00	100.00
11. Fortum Distribution AB / Ellevio	11.26	0.00	9.77	0.00	0.00
12. Hafslund Nett AS	10.91	5.45	4.08	0.00	82.74
13. Iberdrola Distribución Eléctrica S.A.	100.00	95.97	100.00	81.29	58.95
14. Østkraft A/S	4.48	0.00	0.00	0.00	67.69
15. RWE Distribution	60.82	100.00	83.23	100.00	64.34
16. Salzburg AG	7.58	0.00	11.47	50.00	0.00
17. Scottish Power Energy Networks	27.07	4.69	12.36	50.00	0.00
18. Skagerak Nett AS	12.53	13.53	8.45	0.00	0.00
19. SWU Netze	4.24	5.04	7.86	0.00	0.00
20. UK Power Networks	11.22	15.92	12.24	0.00	0.00
21. Vattenfall AB / Gotlands Energi AB	58.91	46.44	34.11	0.00	0.00
22. Western Power Distribution	36.08	4.70	25.65	0.00	0.00
23. Wien Energie Stromnetz	7.95	5.20	4.49	0.00	0.00

4.2 *Second-Stage Results—DSOs' Performance in the SGs Field*

Similarly to the first-stage analysis, weight restrictions were applied in the second stage of the framework. The weights were chosen to reflect JRC's guidelines for SG implementation, as well as the general investment choices made by the majority of the DSOs. The weight restrictions implemented are presented in (6). In (6), u_{SNM}, u_{DSM}, $u_{\text{DG\&S}}$, $u_{\text{e-Mobility}}$ and u_{others} correspond to the pure weights assigned to each SG area (e.g., SNM, DSM, DG&S, e-Mobility and "others"). The parameters $\overline{y}_{\text{SNM}}$, $\overline{y}_{\text{DSM}}$, $\overline{y}_{\text{DG\&S}}$, $\overline{y}_{\text{e-Mobility}}$ and $\overline{y}_{\text{Others}}$ correspond to the average value of the Composite Indicator representing the performance of DSOs in each area considered (SNM, DSM, DG&S, e-Mobility and "others") computed for all DMUS of the sample.

$$\begin{aligned} u_{\text{SNM}}\,\overline{y}_{\text{SNM}} &\geq u_{\text{DSM}}\,\overline{y}_{\text{DSM}} \\ u_{\text{DSM}}\,\overline{y}_{\text{DSM}} &\geq u_{\text{DG\&S}}\,\overline{y}_{\text{DG\&S}} \\ u_{\text{DG\&S}}\,\overline{y}_{\text{DG\&S}} &\geq u_{\text{e-Mobility}}\,\overline{y}_{\text{e-Mobility}} \\ u_{\text{e-Mobility}}\,\overline{y}_{\text{e-Mobility}} &\geq u_{\text{others}}\,\overline{y}_{\text{others}} \end{aligned} \tag{6}$$

Figure 2 illustrates the DSOs' overall performance (CI_{2017}) for the year 2017.

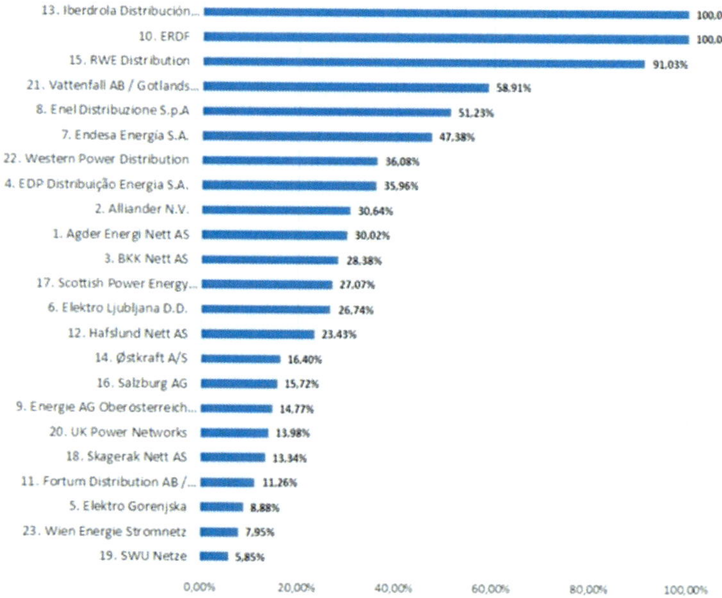

Fig. 2 Second-stage results for year 2017. DSOs' performance in the SG field

Figure 2 shows that Iberdrola, ERDF and RWE are clearly the leaders in terms of performance in the SGs research field. These DSOs made the largest investments in SG projects, in a wide variety of areas. Looking at DSOs with similar levels of investment, e.g. Endesa and Vattenfall, it is possible to gain interesting insights. Until the year 2017, Endesa conducted 19 SG projects with a total investment of 444.65 M€. Vattenfall, on the other hand, conducted only 13 projects, with a total investment of 301.45 M€. However, from Fig. 2 it is possible to see that Vattenfall has a better overall performance than Endesa, 58.91% to 47.38%. This can be explained by the focus of Vattenfall in international projects. As it is possible to see from Tables 1 and 2, all of Vattenfall's projects have been international, 13 (including one as leader), while Endesa has conducted 11. This highlights one of the most interesting aspects of using DEA for performance assessment: the fact that each unit is always shown at its best light. Thus, in the case of these DSOs, the priority given by Vattefall to international projects, as well as the leadership taken in one of this, has contributed to a better overall performance when compared to Endesa.

4.3 Malmquist Index Results—Evolution of DSOs' Performance over Time

To evaluate the evolution of DSOs' performance between the 2013 and 2017 we used a Malmquist Index $\left(\mathrm{MI}^{13-17}\right)$. The same models were used to evaluate the DSOs performance in the years 2013 and 2017, including similar weight restrictions. The results obtained are listed in Table 4. Wien Energie was excluded from the Malmquist Index analysis due to missing data in the JRC database for the year 2013.

For the remaining 22 DSOs considered in our sample, 8 DSOs decreased their performance between 2013 and 2017, whilst 12 increased performance in relation to the peers, and 2 maintained. Note that, in the context of performance evaluations using the Malmquist index, values of $\mathrm{MI}^{t,t+1}$ greater than one signal performance improvements between t and $t+1$, whilst values smaller than one signal decline in performance.

The biggest decrease in performance corresponds to Elektro Gorenjska and biggest increase corresponds to Vattenfall AB/Gotlands Energi AB. From the set of the best performing DSOs in 2017 (Iberdrola, ERDF and RWE) it is possible to see that RWE was the only one which decreased its performance, going from one of the best-performing DSOs, to the third best in 2017.

The MI can be decomposed in two sub-indices (EC and FS), as proposed by [17]. The first sub-index (EC) measures the efficiency change between two time periods. The second sub-index (FS) measures frontier shift between the periods considered. EC evaluates the distance to the frontier t of the DSO under evaluation in t, compared with the distance to the frontier $t+1$ of that DSO in $t+1$. FS compares the distance between the best practice frontiers in t and $t+1$, evaluated at the input-output mix of the DSO under evaluation, both in t and in $t+1$.

Table 4 Malmquist Index results. DSOs performance evolution between 2013 and 2017

DSO	$MI^{13,17}$	CI^{13}	CI^{17}	EC	FS
1. Agder Energi Nett AS	1.07	26.29	30.02	1.14	0.94
2. Alliander N.V.	1.03	29.11	30.64	1.05	0.97
3. BKK Nett AS	1.05	25.84	28.38	1.10	0.95
4. EDP Distribuição Energia S.A.	0.98	37.73	35.96	0.95	1.02
5. Elektro Gorenjska	0.78	14.66	8.88	0.61	1.29
6. Elektro Ljubljana D.D.	1.12	21.24	26.74	1.26	0.89
7. Endesa Energía S.A.	1.08	40.57	47.38	1.17	0.93
8. Enel Distribuzione S.p.A	0.95	57.04	51.23	0.90	1.06
9. Energie AG Oberösterreich Netz	1.15	11.13	14.77	1.33	0.87
10. ERDF	1.00	100.00	100.00	1.00	1.00
11. Fortum Distribution AB / Ellevio	0.97	11.88	11.26	0.95	1.03
12. Hafslund Nett AS	1.02	22.49	23.43	1.04	0.98
13. Iberdrola Distribución Eléctrica S.A.	1.00	100.00	100.00	1.00	1.00
14. Østkraft A/S	1.00	16.27	16.40	1.01	1.00
15. RWE Distribution	0.95	100.00	91.03	0.91	1.05
16. Salzburg AG	1.19	11.13	15.72	1.41	0.84
17. Scottish Power Energy Networks	0.83	39.74	27.07	0.68	1.21
18. Skagerak Nett AS	1.12	10.57	13.34	1.26	0.89
19. SWU Netze	0.90	7.23	5.85	0.81	1.11
20. UK Power Networks	1.05	12.73	13.98	1.10	0.95
21. Vattenfall AB/Gotlands Energi AB	1.27	36.41	58.91	1.62	0.79
22. Western Power Distribution	0.94	40.98	36.08	0.88	1.07
23. Wien Energie Stromnetz	na	na	7.95	na	na
Average	1.02	35.14	34.57	1.05	0.99

The analysis of the average values of the MI and its components (see last line of Table 4) reveals that most of the performance improvements observed in the period 2013–2017 are due to the increased proximity of the DSOs to the best-practice frontier in 2017, relatively to 2013. This is evidenced by a value greater than one of the efficiency change component (EC = 1.05), while the average FS was basically kept unchanged (FS = 0.99), reflecting greater homogeneity in terms of participation in R&D projects.

5 Conclusions

This work describes a benchmarking exercise covering a set of European electricity DSOs regarding their investment in smart grids research projects. This analysis was conducted recurring to the DEA technique.

The methodology applied involved several stages. The first stage is focused on the computation of an indicator per SG area—SNM, DSM, DG&S, e-Mobility and "Others". This last category included the project types that could not fit in any of the other categories specified. The second stage performs an aggregation of the indicators computed in the first stage, in order to provide an overall indicator of the DSOs' performance. The final stage analysed the trends in the evolution of relative performance between 2013 and 2017. In all stages of the analysis, weight restrictions were applied to reflect general decision maker preferences regarding the development of smart grids. The three DSO in the top 3 positions in both years were ERDF, Iberdrola and RWE Distribution. RWE is the only DSO from this set of top performers that decreased its relative performance in this period. Concerning the remaining DSOs, the largest increase in performance occurred for Vattenfall AB/Gotlands Energi AB, with a MI of 1.27, and the highest decrease occurred for Elektro Gorenjska, with an MI index of 0.78.

Overall, it was observed that the efficient frontier was kept unchanged for the analysed period. As a result, although the best-performing DSOs in 2013 contributed significantly to the initial research efforts in the SG field, it was the remaining DSOs which, altogether, contributed to keep the research pace in the field.

References

1. Liu, Y., Bi, R.: Research of fiscal expenditure performance evaluation quantization – Data envelopment analysis application. In: 2009 IEEE/INFORMS International Conference on Service Operations, Logistics and Informatics, pp. 533–536 (2009)
2. Zhou, H., Yang, Y., Chenb, Y., Zhu, J.: Data envelopment analysis application in sustainability: the origins, development and future directions. Eur. J. Oper. Res. **264**(1), 1–16 (2018)
3. Gangale, F., Vasiljevska, J., Covrig, F., Mengolini, A., Fulli, G.: Smart grid projects outlook 2017: facts, figures and trends in Europe, EUR 28614 EN. https://doi.org/10.2760/15583.
4. Charnes, A., Cooper, W.W., Rhodes, E.: Measuring the efficiency of decision making units. Eur. J. Oper. Res. **2**(6), 429–444 (1978)
5. Allen, R., Athanassopoulos, A., Dyson, R.G.: Weights restrictions and value judgements in data envelopment analysis: evolution, development and future directions. Ann. Oper. Res. **73**, 13–34 (1997)
6. Cook, W.D., Tone, K., Zhu, J.: Data envelopment analysis: prior to choosing a model. Omega **44**, 1–4 (2014)
7. Saltelli, A.: Composite indicators between analysis and advocavy. Soc. Indic. Res. **81**(1), 65–77 (2007)
8. Cherchye, L., Moesen, W., Rogge, N., Puyenbroeck, Van T.: An introduction to Benefit of the doubt composite indicators. Soc. Indic. Res. **82**(1), 111–145 (2007)
9. Koopmans, T.: Analysis of production as an efficient combination of activities. In: Activity Analysis of Production and Allocation, pp. 33–97. Wiley, New York (1951)

10. Cook, W., Kress, M.: A data envelopment model for aggregating preference rankings. Manage. Sci. **36**(11), 1302–1310 (1990)
11. Lovell, C.: Measuring the macroeconomic performance of the taiwanese economy. Int. J. Prod. Econ. **39**(1–2), 165–178 (1995)
12. Lovell, C., Pastor, J., Turner, J.: Measuring macroeconomic performance in the OECD: a comparison of european and noneuropean countries. Eur. J. Oper. Res. **87**(3), 507–518 (1995)
13. Zanella, A., Camanho, A., Dias, T.: Undesirable outputs and weighting schemes in composite indicators based on data envelopment analysis. Eur. J. Eur. Res. **245**, 517–530 (2015)
14. Wong, Y.H., Beasley, J.E.: Restricting weight flexibility in data envelopment analysis. J. Oper. Res. Soc. **41**(9), 829–835 (1990)
15. Thompson, R.G., Langemeier, L.N., Lee, C.T., Lee, E., Thrall, R.M.: The role of multiplier bounds in efficiency analysis with application to Kansas farming. J. Econ. **46**(1–2), 93–108 (1990)
16. Fare, R., Grosskopf, S., Norris, M., Zhang, Z.: Productivity growth, technical progress, and efficiency change in industrialized countries. Am. Econ. Rev. **84**(1), 66–83 (1994)
17. Fare, R., Grosskopf, S., Lindgren, B., Roos, P.: Productivity change in Swedish pharmacies 1980–1989 a nonparametric Malmquist approach. J. Prod. Anal. **3**, 85–102 (1992)

Towards an Integrated Decision-Support Framework for the New Generation of Manufacturing Systems

Miguel Vieira, Fábio Coelho, Cátia da Silva, Bruna Mota, Joana Guapo, Rodrigo Macedo, Bruno Gonçalves, Samuel Moniz, Tânia Pinto-Varela, Ana Carvalho, Susana Relvas, and Ana Paula Barbosa-Póvoa

Abstract The global innovation paradigm for industry is setting new research challenges for the management of manufacturing systems. With Industry 4.0 transformation across multiple enterprise levels, advanced automation gained a greater relevance to business intelligence models, as it aims to leverage integrated approaches to improve competitiveness and responsiveness under operational and market uncertainty. In this work, an integrated framework is discussed to develop management solutions for the next generation of industrial manufacturing systems. The underlying idea is to discuss existing challenges for designing and implementing effective decision-support systems. Both modeling and decision-support perspectives are provided so as to develop a framework capable to address reactive production planning and scheduling problems supported by efficient in-house logistics systems, while pursuing operational sustainability targets.

Keywords Industry 4.0 · Planning · Scheduling · In-house logistics · Sustainability · Decision-support framework

1 Introduction

The management of the operational complexity poses significant challenges to the implementation of effective decision support systems. The dynamic structure of

M. Vieira · F. Coelho · C. da Silva · B. Mota · J. Guapo · R. Macedo · T. Pinto-Varela (✉) · A. Carvalho · S. Relvas · A. P. Barbosa-Póvoa
CEG-IST, Instituto Superior Técnico Universidade de Lisboa, Lisboa, Portugal
e-mail: tania.pinto.varela@tecnico.ulisboa.pt

A. P. Barbosa-Póvoa
e-mail: apovoa@tecnico.ulisboa.pt

B. Gonçalves
Departamento de Produção e Sistemas, Universidade do Minho, Guimarães, Portugal

M. Vieira · S. Moniz
Univ Coimbra, CEMMPRE, Departamento de Engenharia Mecânica, Universidade de Coimbra, Coimbra, Portugal

S. Relvas et al. (eds.), *Operational Research*, Springer Proceedings in Mathematics & Statistics 374, https://doi.org/10.1007/978-3-030-85476-8_14

most manufacturing processes requires the simultaneous assessment of different business criteria for the majority of strategic and operational decisions [1]. With the advent of digitalization, many manufacturing systems have been focusing on the improvement of operational efficiency with the use of process automation technologies and data science. The goal is to potentiate an enhanced and on-line control of the whole system (e.g. work centers, production lines, or plants). Along with this trend to move to a comprehensive digitalization of the decision-support processes, other Industry 4.0 technologies are becoming more relevant. In short, 'Industry 4.0' has been recognized as a popular worldwide initiative to promote production and process innovation, joining efforts of academia and industry towards the development of advanced sensors, collaborative automation and integrated analytical tools for the new generation of smart productive systems [2]. This new wave of process automation defines an updated paradigm for industrial decision-support processes, considering the design of production layouts, the human-machine interaction, the scheduling/planning optimization of highly adaptive manufacturing processes, the efficient resources management, among others.

The expected benefits of this technological transformation are huge and have the potential to leverage the industrial process into superior levels of efficiency [3]. However, the adoption of Industry 4.0 paradigm entails substantial and intricate changes in decision-support systems. Two of the most relevant are systems full integration and a shorter time delay between decisions. Achieving this will increase the speed at which decisions are made and will potentiate better (or even optimal) decision-support. Therefore, the new generation of automated manufacturing systems needs to be adaptive and demands (near) real-time decision-support. It is clear that the advances on setting the optimal levels of automation and keeping the right levels of human-centered automation will play an important role, since these provide the basis of data and structure for building robust models (optimization and simulation) of whole systems. Withal, this automation trend is also raising additional concerns on how to manage the correct balance over multiple sustainability issues. These challenging production environments have enabled the development of advanced planning and scheduling methodologies, following the digital transformation of processes. Within this research scope, the present work aims at the development of a framework for production planning and scheduling solutions supported by efficient in-house logistics operations where sustainability goals are pursued.

The remainder of the proposed work follows the need for integrated decision-support in Sect. 2, and the suggested decision-support framework in Sect. 3, with the description of the decisional scope with focus on the methodologies to approach them. The last section refers to the conclusions and future research directions.

2 The Need for Integrated Decision-Support

The competitiveness of manufacturing industries depends largely on its own productivity, flexibility, and agility to react to market uncertainty. Despite many production lines and plants are already supported by recent technologies (e.g. software architectures that provide comprehensive connectivity of data flows), recent technological trends such as manufacturing/process automation and collaborative robotic resources are expected to leverage the global efficiency of the systems [4, 5]. This transformation will have impact not only at improving the flexibility and agility of industrial production, but will bring additional degrees of decentralization to decision-support. Moreover, it represents a challenge for the traditional modelling techniques, where resources are mostly represented with static capacity and range of flexibility not totally explored. More often than not, both academic advances and industrial cases are still limited in demonstrating the real value against the expectations [6]. With the focus at the decision-support level, tools will have to evolve to deal with production technologies with high flexibility, capable of controlling a large set of variables (at different scales) and provide solutions for multiple integrated problems. Therefore, the main motivation scope relies on an integrated model-based framework for decision-support on the management of the new generation of production systems, while fostering the efficient use of advanced manufacturing technologies. These novel systems will require to encompass the tradeoffs at different operational dimensions, with particular importance to the sustainability implications of the decision-support process at the economic, environmental and social levels.

The operational research field applied to Industry 4.0 must evolve to comprise the technological advances in digitalization and automation applied to manufacturing environments. This will expand its application to lever the majority of industrial drivers, establishing an optimized digital value chain to potentiate simultaneous planning of products and processes improvement on quality and lead time indicators. As noted by Hofmann and Rüsch [7], some of the key components consider:

(i) Cyber-physical systems, to integrate computers, networks monitoring and process control in the same infrastructure;
(ii) Internet of Things (IoT), to enable a communication platform among the different levels, enabling an end-to-end integration of the value chain;
(iii) Internet of Services (IoS), based on the idea that customized services are made easily available to market through web technologies;
(iv) Smart factory, which defines the interaction of the previous components, built on the idea of a decentralized manufacturing system where products, machinery, transport systems and workers are optimized resources in a highly flexible environment.

Beyond these drivers, high tech wearables, human-collaborative robotics, augmented reality applications, big data analytics and cloud computing are usually seen in a straightforward horizontal and vertical integration of the value network, not only at all the company departments but also at customers and suppliers' sides. Therefore, the goal of a Smart factory, which can cover from the product design stage to the shipping of products, is to have an integrated system where different production and resources are interconnected to improve and make more efficient production systems [8].

The new systems need novel planning and scheduling methodologies to cope with the increasing process complexity. Integrated manufacturing systems are considered as an industrial environment developed for the novel generation of automated process technologies [9]. As noted by Dörmer et al. [10], it impacts the overall engineering design, process planning, manufacturing resources allocation, quality management, and storage retrieval functions, where improving the quality of the decisions in those fields gives rise to complex combinatorial optimization problems. The industrial requirements can address operational scenarios for plant design or process layouts, allowing to study different buffers capacity, bottlenecks workstations or routes for logistics tasks. With regard to the future of in-house logistics, concepts of intelligent warehousing with autonomous and based systems of production sensors increase can potentiate the flexibility on how objects are transported, stored, supplied or produced. But, as noted by Nielsen et al. [11], there is little research in real-world applications of autonomous mobile robots. For example, the order picking activity continues to be made by humans, since until now most automatic systems have failed to respond to the prerequisites of safety, speed and precision, which are essential features to both service level and workspace collaboration with humans. Nevertheless, the research community is striving with the development of collaborative robotic systems that physically interact with humans in a shared workspace [7]. But the impact of this technological transformation is in fact much broader and influences the overall sustainability of these processes. Accounting the growing importance of sustainable operations, as noted by Barbosa-Póvoa et al. [12], it is necessary to integrate it into the manufacturing industrial management, bridging the economic expectations with social and environmental concerns.

In short, advanced manufacturing systems will require methodologies built on more agile, autonomous, and decentralized solutions. In what concerns optimization techniques, despite the relevance of traditional mathematical models, efficient approaches are becoming key to address the coherence of the different decisional levels, as reviewed by Gen and Lin [13]. For example, regarding planning and scheduling solutions, can be verified that most traditional optimization approaches spend more effort in the modeling aspects of the problem and in the development of methods to deal with the computational burden, while novel data-driven approaches attempt primarily to make a better use of the existing process data to solve the problem. These authors suggest that there has been a perspective shift in the literature from manufacturing system design to manufacturing system operation, with the use of advanced data interconnectivity to integrate related operations towards an enhanced business performance.

3 Integrated Framework

The development of an efficient model-based decision-support framework arises from the increasing complexities in operational problems to integrate different decisional levels. As reviewed by Power and Sharda (2007) [14], these systems involve a human-computer interaction with the knowledge/experience held by the process stakeholders, structured in a way that a model-based methodology provides the information to managers decide the best solution for a complex problem. The implementation of such framework must follow several criteria, as suggested by Vieira et al. [15], namely:

(i) the modeling approach should be flexible and generic enough to integrate several information sources;
(ii) an effective user interface must be provided for the control of inputs, data processing, and graphical KPI visualization;
(iii) the solution must be robust, scalable, and stable, allowing the generation of reactive solutions to process disturbances; and
(iv) the platform must be easily integrated and maintainable within industrial software platforms (e.g. ERP-software and plant control systems).

The framework is structured around the proposed solution approach that combines design, planning, and operational sustainability decision-support, as can be seen in Fig. 1. It should enable the management of production planning/scheduling functions while considering efficient in-house logistics operations and operational sustainability. The assessment of the different solutions is constrained by the strategic criteria, automation level and solution flexibility, while considering the manufacturing structure attributes as inputs. The evaluation of the outputs requires KPIs definition, validation and implementation control.

Despite the increased effort made by the research community, it can be acknowledged that the impact of these technological advances in operational decision-support

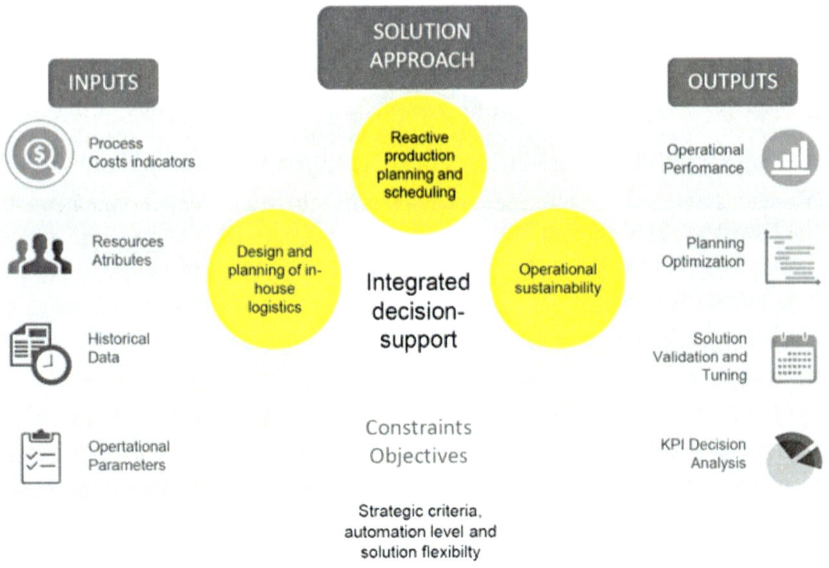

Fig. 1 Integrated decisional framework

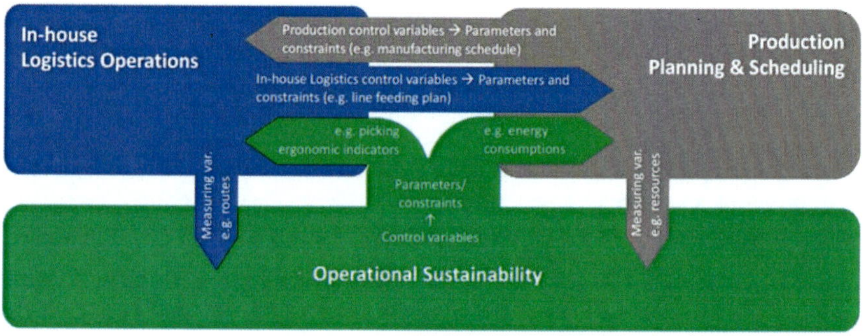

Fig. 2 Multilevel modeling and decision-support in the manufacturing systems

strategies is still an open question. Relevant works in this field are limited in the sense that, to our knowledge, they show short-sighted interconnectivity amid the three topics. A few examples of these works are the cases of Boysen and Emde [16] and Emde and Gendreau [17] who explored scheduling in-house logistics supply to assembly lines, Stock and Selinger [18] reviewed the sustainability opportunities in retrofit manufacturing operations, Wang et al. [19] proposed a multiple criteria approach for sustainable process planning and scheduling optimization, and Chen et al. [20] and Tsarouchi et al. [21] explored the issues of human–robot interaction regarding social metrics in task allocation and scheduling.

The methodological structure for the integrated solution approach is defined in Fig. 2, considering the framework focused on reactive production planning and scheduling problems, supported by efficient in-house logistics operations and assessed by operational sustainability indicators. The iterative information flow within the three components considers the control variables and measures of the decisions taken at each component as input parameters or constraints to the others, converging to a solution that combines the requirements from production and in-house logistics with suitable operational sustainability indicators.

The main areas to be explored and insights on how to explore each component of the framework are explained in the following sections.

3.1 Production Planning and Scheduling

In a context of demand variability, resources shortages, maintenance breakdowns and limited production capacity availability, reactive planning and scheduling solutions are required. By definition, planning is typically associated with long/medium-term aggregated decisions of production activities. This, in turn, requires efficient scheduling to perform the short-time allocation of resources and sequencing of tasks over time periods, which often implies a combination of both in a detailed problem solution [22]. Despite the significant developments in exact/non-exact modeling techniques in recent years, it is acknowledged that generalized mathematical formulations are limited by the inherent computational hindrance with the system modeling growing in combinatorial complexity [23]. Exact methods can easily solve small-sized scheduling problems but show significant combinatorial complexity, since most scheduling problems are $\mathcal{NP}$-hard, where decomposition techniques are gathering relevance. Advanced metaheuristics can combine high-level structures for developing specific heuristic methods to fit a particular kind of problem with efficient computational time, although unable to guarantee the global optimal solution [9]. Simultaneously, the development of novel simulation platforms has empowered the evaluation of scenarios that mimic real operations and explore the complexity of nonlinear and stochastic systems [24]. The potential hybridization of these methodologies has been gathering significant attention by combining their strengths to address the solution search-space. As noted by Figueira and Almada-Lobo [25], simulation and optimization approaches combine the advantages of the process detail representation with the ability to optimize solutions. With different abstraction levels in each model, the simulation includes features not addressed in the optimization to decrease complexity, while considering iterative parameters adjustment to allow solution convergence. The availability of on-line data on operational parameters also brought new possibilities for dynamic scheduling modeling strategies, able to plan the tasks and resources allocation facing potential operational disruptions. Solutions exploring reactive responses to unexpected events with low computation burden can be modeled with event driven policies coupled with exact formulation and meta-heuristic approaches where uncertainty is to be considered.

The methodology approach to the problem of production planning and scheduling, according to Moniz et al. [26], which is now extended to integrate the assessment of efficient in-house logistics operations where sustainability goals are pursued, can

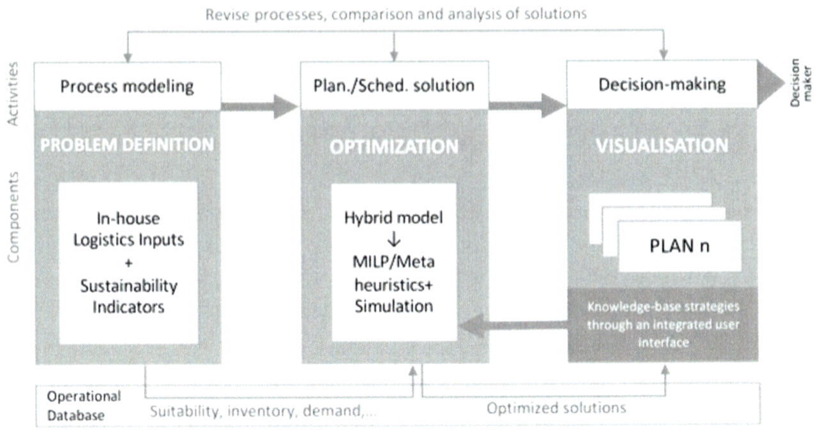

Fig. 3 Methodology structure for the planning and scheduling decision-support (adapted from Vieira et al. [15])

be described in three main steps, as depicted in Fig. 3:

- the definition of the problem structure, providing the support components for the hybrid optimization/simulation model;
- the modeling approach generating the solutions set for the given problem able to efficiently react to changes;
- the analysis of solutions via a user-interface providing the interactive procedure to support the decision-maker.

The development of the hybrid methodology provides significant advantages to address the complexity of these industrial problems [27]. Take the example of a case in which the goal is to determine the optimal allocation of automated shared resources to perform a production mix per demand period in a multistage assembly line, considering the cost-efficient use of resources. This hybrid approach enables to generate a solution at the planning capacity level (e.g. with a MILP or metaheuristic model) that is, iteratively, evaluated at the scheduling level with the visualization of the dynamic behavior of the production process modeled in a simulation software. Besides the production mix, the capacities and workload of workstations in the manufacturing system are affected by various factors such as buffer capacity, task sequencing, operational uncertainty, input policy or dispatching rules. The main components available at the simulation level are used to achieve the most accurate representation of the manufacturing system, which requires advanced programming for modeling specific process features. Based on simulation and optimization methods, the proposed framework addresses the core planning/scheduling industrial problem, evaluating the solution convergence with demand and operational uncertainty as a reactive strategy. The inputs from the other two components through their methods will support the solution interaction, by analysis of in-house logistics operations and the assessment of operational sustainability indicators, providing an integrated solution approach for these advanced manufacturing systems able to react efficiently and timely to changes of the production conditions [28].

3.1.1 Design and Planning of In-House Logistics

Product based manufacturing industries rely on strong in-house logistics operations, such as assembly line feeding, parts distribution and supermarket management (decentralized warehouses), which support efficient production strategies. In this set, inventory accuracy is a keystone of production efficiency, and prompt decision making in flexible manufacturing raises the need for optimized shop-floor operations. The challenge lies in the integration of the decisions made at strategic management levels, such as the warehouse/supermarket design and layout, in solutions for in-house logistics operations. Decision-support tools need to integrate all shop-floor operations, such as storing/sequencing of parts, supermarket and inventory management. The methodology proposed defines a solution approach using simulation-based methods, built on discrete-event simulation platforms, such as object-oriented sim-

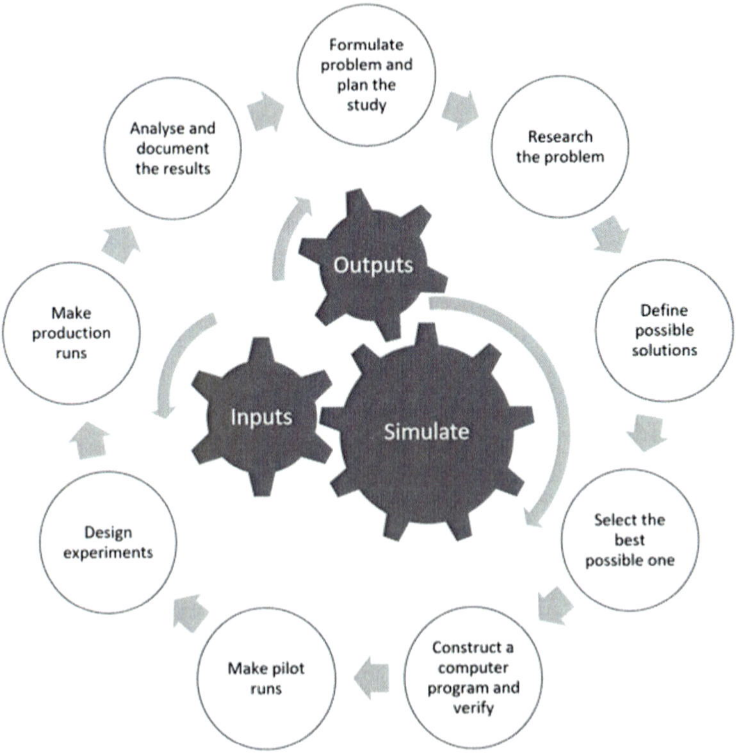

Fig. 4 Methodology structure for in-house logistics operations assessment

ulation software, allowing the flexibility to evaluate the performance of different assembly line feeding scenarios (Fig. 4). Particular relevance is given to lead times estimation, comprised of both processing and queuing time, which can be used to support the planning/scheduling solution. For example, a detailed simulation model enables to validate and verify the line feeding process using different resources, the actual moment of jobs in a time-ordered sequence and the visual animation of the system behavior throughout the plan.

With a focus on supermarket management, to provide another example, there is a consensus in the literature that this concept is little explored by the research community, as discussed by Emde and Boysen [29] and Saaidia et al. [30]. The works that address assembly line feeding are tipically focused on the characteristics of routes, scheduling of tow-trains, or size and location of these logistics supermarkets [31]. Although there are several studies that address the problems related to supermarket and picking policies [32–38], these mostly rely on optimization models to the detriment of the advantages of simulation models. By proposing a simulation-based model approach it is possible to test real size logistic supermarkets, with fleets of robots and, therefore, test the viability of the developed algorithms in full production systems. Moreover, this contributes to the automation of internal logistics activities to improve competitiveness and support the company's value-added integration of the sustainability assessment. In the scope of multi-level decision-support, simulation models can provide useful data information and indicators (e.g. cycle time, makespan, sort of resources, capacity utilization), by allowing to introduce uncertainty and evaluate how the system reacts in different scenarios, towards having a "digital twin" of the production system.

3.1.2 Operational Sustainability

New manufacturing systems arising from Industry 4.0 are bound to result in significant productivity and efficiency-related improvements [39, 40] but they are also expected to introduce new challenges, particularly when it comes to balancing the sustainability of all the involved stakeholders. Undoubtedly, one of the main lead factors of Industry 4.0 is the increased data availability of processes. Data availability has always been one of the main setbacks underlying sustainability assessment, which has contributed to the suboptimal performance of many existent methodologies. In this new industrial era, such data availability is expected to improve the transparency and control of the economic, environmental and social dimensions of sustainability, from strategic to operational decisional levels. A noble example relies in the Global Reporting Initiative [41], the best-known framework for voluntary reporting of environmental and social performance by business and other organizations worldwide. It is therefore important to develop decision support tools to take it into account, by introducing adequate and meaningful sustainability indicators to address the tradeoffs and explore the possibilities of new "win-win-win" solutions. Based on the information available from these resources and supported by ongoing industrial field studies, in Table 1 a broad set of indicators can be projected for the sustainability assessment of production planning and scheduling activities in an Industry 4.0 manufacturing context. Regarding the economic dimension of sustainability, Industry 4.0 is expected to significantly decrease unitary production costs and, consequently, facilitate revenue and profit increase. This can be achieved through a number of different mechanisms, either through increased quality rates, increased productivity rates, decreased cycle times, or even decreased raw material consumption. As a trade-off, it is required substantial investment to adapt the manufacturing activities to the technological context. These indicators are directly related to the environmental dimension of sustainability, since increased productivity, efficiency and quality are also potentially associated with decreased waste generation and raw material consumption, increased recycling rates (with control over the quality and composition of generated waste, non-conforming products or even end-of-life products), and reduction in water consumption and greenhouse gases emissions (assuming the same demand levels). Regarding the social dimension, three main aspects can be identified in terms of sustainability within the planning/scheduling decision context: worker health and safety, worker labor conditions and employment, and worker satisfaction. Focusing on worker health and safety, possible indicators are related to ergonomics,

Table 1 Sustainability indicators suitable at the production planning and scheduling decision-support level in an Industry 4.0 manufacturing context

	Sustainability indicators
Economic dimension	• Profit • Investment • Revenue • Cost (total or unit) • Cycle time • Productivity • Quality
Environmental dimension	• Energy consumption • Water consumption • Raw material extraction • Waste generation • Recycling rates • Greenhouse gases emissions
Social dimension	• Ergonomic indicators • Injury frequency • Accident frequency • Accidents with irreversible consequences • Fatalities per employee • Layoffs • Employee turnover • Training man days • Absenteeism rate

injury and accident frequency and severity. A good example of the importance of these indicators in Industry 4.0 context is related to collaborative robots, which are intended to collaborate with humans in assembly lines, improving safety and reducing the physical burden on the human co-workers, while also improving productivity and quality rates [42]. On the other hand, one of the main challenges associated with manufacturing automation is the impact in human employment. It is important to incorporate additional indicators capable of assessing and incorporating efforts towards avoiding layoffs, as are training man hours. Finally, employee satisfaction is also a very important aspect of these systems since unsatisfied workers can hinder productivity levels. Absenteeism rates have been identified as indirect indicators of this aspect [43]. Overall, it is important to incorporate a solid set of sustainability indicators, capable of providing a clear image of both the challenges and opportunities associated with each of the potential solutions arising from the different decisional levels discussed in this work as an integrated management framework.

4 Conclusions

Following the new trends in industrial processes, the next generations of production systems require an integrated approach that combines different methodologies for the optimization of production planning and scheduling supported by the management of efficient in-house logistics operations. Sustainability assessment needs to be present at both of these steps, so as to guaranty the economic, environmental and social viability of the obtained solutions. The core of this work lies in underlying the necessary steps towards the development of such an approach so as to enable the assessment of the flexibility, robustness and sustainability of solutions. A multimethod framework is then proposed to assist decision makers in the tradeoff evaluations of its various cross-functional dimensions. The integration of different decisional contexts resumes the challenges posed by these novel manufacturing systems. Different methodological approaches, based on OR methods, are essential to guaranty the operationality of the production planning and scheduling solutions in a sustainable way.

The implementation challenges of the framework must evolve to consider the solution application to uncertainty contexts, by combining the different methodologies to potentiate optimized management decisions for the upcoming eras in the manufacturing industry.

Acknowledgements This work is part of a broader research project named DM4Manufacturing (Project website: https://dm4manufacturing.inesctec.pt/), which aims at developing multidisciplinary manufacturing decision-support tools aligned with the efficient use of automation technologies, able to perform an optimized balance between design and operation decisions, particularly in high-mix and high-customization manufacturing systems. Therefore, the authors would like to acknowledge the financial support of Fundação para a Ciência e Tecnologia from FEDER through the POCI program under the project POCI-01-0145-FEDER-016418 and under the project UIDB/00285/2020.

References

1. Varma, V.A., Reklaitis, G.V., Blau, G.E., Pekny, J.F.: Enterprise-wide modeling & optimization - An overview of emerging research challenges and opportunities. Comput. Chem. Eng. **31**(5), 692–711 (2007)
2. Oesterreich, T.D., Teuteberg, F.: Understanding the implications of digitisation and automation in the context of industry 4.0: a triangulation approach and elements of a research agenda for the construction industry. Comput. Ind. **83**, 121–139 (2016)
3. Bauer, C., Wee, D.: Manufacturing's next act. McKinsey Operations Insights (2015)
4. Isaksson, A.J., Harjunkoski, I., Sand, G.: The impact of digitalization on the future of control and operations. Comput. Chem. Eng. **114**, 122–129 (2018)
5. Helo, P., Phuong, D., Hao, Y.: Cloud manufacturing–Scheduling as a service for sheet metal manufacturing. Comput. Oper. Res. (2018)
6. Lu, Y.: Industry 4.0: a survey on technologies, applications and open research issues. J. Ind. Inf. Integr. **6**, 1–10 (2017)
7. Hofmann, E., Rüsch, M.: Industry 4.0 and the current status as well as future prospects on logistics. Comput. Ind. **89**, 23–34 (2017)

8. Li, Q., Tang, Q., Chan, I., Wei, H., Pum, Y., Jiang, H., Zhou, J.: Smart manufacturing standardization: architectures, reference models and standards framework. Comput. Ind. **101**, 91–106 (2018)
9. Gen, M., Lin, L., Zhang, H.: Evolutionary techniques for optimization problems in integrated manufacturing system: state-of-the-art-survey. Comput. Ind. Eng. **56**(3), 779–808 (2009)
10. Dörmer, J., Günther, H.O., Gujjula, R.: Master production scheduling and sequencing at mixed-model assembly lines in the automotive industry. Flex. Serv. Manuf. J. **27**(1), 1–29 (2015)
11. Nielsen, I., Dang, Q.-V., Bocewicz, G., Banaszak, Z.: A methodology for implementation of mobile robot in adaptive manufacturing environments. J. Intell. Manuf. **28**(5), 1171–1188 (2017)
12. Barbosa-Póvoa, A.P., da Silva, C., Carvalho, A.: Opportunities and challenges in sustainable supply chain: an operations research perspective. Eur. J. Oper. Res. **268**(2), 399–431 (2018)
13. Gen, M., Lin, L.: Multiobjective evolutionary algorithm for manufacturing scheduling problems: state-of-the-art survey. J. Intell. Manuf. **25**(5), 849–866 (2014)
14. Power, D.J., Sharda, R.: Model-driven decision support systems: concepts and research directions. Decis. Support Sys. **43**(3), 1044–1061 (2007)
15. Vieira, M., Pinto-Varela, T., Barbosa-Póvoa, A.P.: A model-based decision support framework for the optimisation of production planning in the biopharmaceutical industry. Comput. Ind. Eng. **129**, 354–367 (2019)
16. Boysen, N., Emde, S.: Scheduling the part supply of mixed-model assembly lines in line-integrated supermarkets. Eur. J. Oper. Res. **239**(3), 820–829 (2014)
17. Emde, S., Gendreau, M.: Scheduling in-house transport vehicles to feed parts to automotive assembly lines. Eur. J. Oper. Res. **260**(1), 255–267 (2017)
18. Stock, T., Seliger, G.: Opportunities of sustainable manufacturing in industry 4.0. Procedia Cirp **40**, 536–541 (2016)
19. Wang, S., Lu, X., Li, X.X., Li, W.D.: A systematic approach of process planning and scheduling optimization for sustainable machining. J. Clean. Prod. **87**, 914–929 (2015)
20. Chen F., Sekiyama K., Cannella F., Fukuda T.: Optimal subtask allocation for human and robot collaboration within hybrid assembly system. IEEE Trans. Autom. Sci. Eng. (2014)
21. Tsarouchi, P., Makris, S., Chryssolouris, G.: Human-robot interaction review and challenges on task planning and programming. Int. J. Comput. Integ. M. **29**(8), 916–931 (2016)
22. Kallestrup, K.B., Lynge, L.H., Akkerman, R., Oddsdottir, T.A.: Decision support in hierarchical planning systems: the case of procurement planning in oil refining industries. Decis. Support Sys. **68**, 49–63 (2014)
23. Harjunkoski, I.C., Maravelias, P.B., Castro, P., Engell, S., Grossmann, I.E., Hooker, J., Méndez, C., Sand, G., Wassick, J.: Scope for industrial applications of production scheduling models and solution methods. Comput. Chem. Eng. **62**, 161–193 (2014)
24. Castro, P.M., Aguirre, A.M., Zeballos, L.J., Méndez, C.A.: Hybrid mathematical programming discrete-event simulation approach for large-scale scheduling problems. Ind. Eng. Chem. Res. **50**(18), 10665–10680 (2011)
25. Figueira, G., Almada-Lobo, B.: Hybrid simulation–optimization methods: a taxonomy and discussion. Simul. Model, Pract. Th. **46**, 118–134 (2014)
26. Moniz, S., Barbosa-Póvoa, A.P., de Sousa, J.P., Duarte, P.: Solution methodology for scheduling problems in batch plants. Ind. Eng. Chem. Res. **53**(49), 19265–19281 (2014)
27. Kim, S.H., Lee, Y.H.: Synchronized production planning and scheduling in semiconductor fabrication. Compu. Ind. Eng. **96**, 72–85 (2016)
28. Jin, M., Tang, R., Ji, Y., Liu, F., Gao, L., Huisingh, D.: Impact of advanced manufacturing on sustainability: an overview of the special volume on advanced manufacturing for sustainability and low fossil carbon emissions. J. Clean. Prod. **161**, 69–74 (2017)
29. Emde, S., Boysen, N.: Optimally routing and scheduling tow trains for JIT-supply of mixed-model assembly lines. Eur. J. Oper. Res. **217**, 287–299 (2012)
30. Saaidia, M., Durieux, S., Caux, C.: A survey on supermarket concept for just-in-time part supply of mixed model assembly lines. In: Proceedings of the 10th International Conference of Modeling and Simulation (2014)

31. Emde, S.: Scheduling the replenishment of just-in-time supermarkets in assembly plants. OR Spectr. **39**, 321–345 (2017)
32. Alnahhal, M., Noche, B.: A genetic algorithm for supermarket location problem. Assembly Autom. **35**(1), 122–127 (2015)
33. Golz, J., Gujjula, R., Gunther, H.-O., Rinderer, S., Ziegler, M.: Part feeding at high-variant mixed-model assembly lines. Flex. Serv. Manuf. J. **24**(2), 119–141 (2012)
34. Faccio, M., Gamberi, M., Persona, A., Regattieri, A., Sgarbossa, F.: Design and simulation of assembly line feeding systems in the automotive sector using supermarket, kanbans and tow trains: a general framework. J. Manag. Control **24**(2), 187–208 (2013)
35. Emde, S., Gendreau, M.: Scheduling in-house transport vehicles to feed parts to automotive assembly lines. Eur. J. Oper. Res. **260**, 255–267 (2017)
36. Battini, D., Calzavara, M., Otto, A., Sgarbossa, F.: The integrated assembly line balancing and parts feeding problem with ergonomics considerations. IFAC-PapersOnLine **49**(12), 191–196 (2016)
37. Caputo, A.C., Pelagagge, P.M., Salini, P.: A decision model for selecting parts feeding policies in assembly liness. Ind. Manage. Data Syst. **115**(6), 974–1003 (2015)
38. Hanson, R., Medbo, L., Johansson, M.I.: Order batching and time efficiency in kit preparation. Assembly Autom. **35**(1), 143–148 (2015)
39. Kagermann, H., Lukas, W.-D., Wahlster, W.: Industrie 4.0: Mit dem Internet der Dinge auf dem Weg zur 4. industriellen Revolution. VDI Nachrichten (2011)
40. Kagermann, H., Wahlster, W., Helbig, J.: Umsetzungsempfehlungen für das Zukunftsprojekt Industrie 4.0 - Abschlussbericht des Arbeitskreises Industrie 4.0. Bmbf.De (2013)
41. GRI (Global Reporting Initiative): GRI Standards Resources, Boston (2018)
42. Cherubini, A., Passama, R., Crosnier, A., Lasnier, A., Fraisse, P.: Collaborative manufacturing wit physical human-robot interaction. Robot. Cim. Int. Manuf. **40**, 1–13 (2016)
43. Popovic, T., Kraslawski, A., Barbosa Póvoa, A.P., Carvalho, A.: Quantitative indicators for social sustainability assessment of supply chains. J. Clean. Prod. **180**, 748–768 (2018)